Grain-based Foods: Processing, Properties, and Heath Attributes

Grain-based Foods: Processing, Properties, and Heath Attributes

Special Issue Editors

Emanuele Zannini
Raffaella Di Cagno

MDPI • Basel • Beijing • Wuhan • Barcelona • Belgrade

MDPI

Special Issue Editors

Emanuele Zannini
University College Cork
Ireland

Raffaella Di Cagno
Faculty of Science and Technology, Free University of Bolzano
Italy

Editorial Office
MDPI
St. Alban-Anlage 66
Basel, Switzerland

This is a reprint of articles from the Special Issue published online in the open access journal *Foods* (ISSN 2304-8158) from 2017 to 2018 (available at: http://www.mdpi.com/journal/foods/special_issues/Cereal-Based_Foods_Processing_Properties)

For citation purposes, cite each article independently as indicated on the article page online and as indicated below:

LastName, A.A.; LastName, B.B.; LastName, C.C. Article Title. *Journal Name* **Year**, *Article Number, Page Range.*

ISBN 978-3-03897-218-1 (Pbk)
ISBN 978-3-03897-219-8 (PDF)

Contents

About the Special Issue Editors

Emanuele Zannini (MSc, Agr., PhD, PhD) finished his PhD in Biomolecular Sciences in 2007 and then worked as an associate professor; since 2009 he has been a senior scientist at the School of Food and Nutritional Sciences, where he also completed a second PhD in Food Sciences. His research activity focus is mainly on food microbiology and food technologies with particular focus on "personalised nutrition" aspects. Dr Zannini has an international reputation for the excellence of his research (H factor: 30—Google Scholar). He has published over 100 peer-reviewed papers, one book, five book chapters and over 25 other papers.

Raffaella Di Cagno (MSc, PhD) has been an associate professor of Food Microbiology since 2015 and began her service at the Free University of Bolzano the January 2017. She received her PhD degree at the School Advanced Studies of Pisa S. Anna. Her main research area is food microbiology with expertise on the molecular microbiology and biotechnology of sourdough, cheese and vegetable/fruit lactic acid bacteria. She is the co-author of 154 articles, published in international journals, which deal with the food microbiology. As recently reported by Scopus, her reviewed publications have been cited ca. 5300 times, with an "H" index of 43.

Preface to "Grain-based Foods: Processing, Properties, and Heath Attributes"

Grain-based food plays a pivotal role in the diet of a vast proportion of the world's population. It therefore has a strong impact on the nutritional quality of the meals consumed by a large number of people and consequently on their health. Additionally, as a result of the rapidly growing global population and limited agricultural area, a sufficient supply of cereals for food has become increasingly challenging. The purpose of this Special Issue is to provide scientific evidence and an overview of the nutritional and health aspects of grain-based products. In more detail, the nutritional and health properties of grain-based products can be strategically improved through a careful selection of raw materials (cultivars); the application of tailored fermentation biotechnology to address specific medical conditions; and (bio)fortification and product reformulation. It also describes the safety aspects of post-harvest preservation and the physical post-harvest treatments able to avoid grain quality deterioration as a limiting factor of the delivery of high quality raw material able to address consumer demand for healthy and nutritious foods.

Emanuele Zannini, Raffaella Di Cagno
Special Issue Editors

foods

MDPI

Article

The Effect of Astaxanthin-Rich Microalgae "Haematococcus pluvialis" and Wholemeal Flours Incorporation in Improving the Physical and Functional Properties of Cookies

A. K. M. Mofasser Hossain [1,2], Margaret A. Brennan [1], Susan L. Mason [1], Xinbo Guo [3], Xin An Zeng [3] and Charles S. Brennan [1,2,*]

[1]   Centre for Food Research and Innovation, Department of Wine, Food and Molecular Biosciences, Lincoln University, Lincoln 7647, New Zealand; AKMMofasser.Hossain@lincolnuni.ac.nz (A.K.M.M.H.); Margaret.Brennan@Lincoln.ac.nz (M.A.B.); Sue.mason@lincoln.ac.nz (S.L.M.)
[2]   Riddet Institute, Palmerston North 4442, New Zealand
[3]   School of Food Science and Engineering, South China University of Technology, Guangzhou 510640, China; xbg720@gmail.com (X.G.); xazeng@scut.edu.cn (X.A.Z.)
*   Correspondence: charles.brennan@lincoln.ac.nz; Tel.: +64-3-4230-637

Received: 23 May 2017; Accepted: 21 July 2017; Published: 26 July 2017

Abstract: Marine-based food supplements can improve human nutrition. In an effort to modulate glycaemic response and enhance nutritional aspects, marine-derived algal food rich in astaxanthin was used in the formulation of a model food (wholemeal cookie). Astaxanthin substitution of cookies made from three flours (wheat, barley and oat) demonstrated a significant reduction in the rate of glucose released during in vitro digestion together with an increase in the total phenolic content (TPC) and antioxidant capacity of the food. The significantly ($p < 0.005$) lower free glucose release was observed from cookies with 15% astaxanthin, followed by 10% and then 5% astaxanthin in comparison with control cookies of each flour. Total phenolic content, DPPH radical scavenging and Oxygen Radical Absorbance Capacity (ORAC) value also notably increased with increase in astaxanthin content. The results evidence the potential use of microalgae to enhance the bioactive compounds and lower the glycaemic response of wholemeal flour cookie.

Keywords: microalgae; *Hematococcus pluvialis*; astaxanthin; bakery products; glycaemic response; antioxidant

1. Introduction

Whole-grains such as wheat, barley and oat make a substantial contribution to our diet. They contain a significant amount of bioactive compounds such as fibre, minerals, vitamins and phytochemicals [1,2] and as such mayplay a major role in enhancing human health by reducing the risk of diabetes [3,4] and cancer [5], while also regulating serum cholesterol [6] and stimulating beneficial gut microbiota [7]. In recent years there has been an increased interest in the utilisation of whole-grain food materials as well as fibre rich ingredients, in cereal products, including bread [8], extruded snack products [9,10], and pasta [11,12]. These pieces of research have investigated the impact of wholegrains and fibre on both the physicochemical characteristics of cereal food products as well as their nutritional quality. A recent review on this subject illustrated that the incorporation of fibre rich ingredients into cereal products often results in negative consumer acceptability [13]. There therefore remains a challenge to both utilise wholegrain cereal products as well as functional food ingredients such a fibre rich materials, into mainstay food products.

Recent research into functional food ingredients has shown an interest in the development of foods containing seaweed or algal materials [14,15]. These materials have been part of the human diet since 600 BC [16] and they have a role of diet in sustaining human due to their diverse range of nutrients and bioactive compounds; such as polysaccharides, proteins, polyunsaturated fatty acids, minerals and significant amounts of antioxidants [17,18]. One such material is *Haematococcus pluvialis*, a single-cell microalgal strain, which is rich source of astaxanthin (10,000–40,000 mg/kg) and associated bioactive ingredients including dietary fibre [19]. Several cell culture and animal studies have reported that astaxanthin has potent antioxidant activity 10 times higher than other carotenoids such as β-carotene, lutein, and zeaxanthin, and 500 times higher than vitamin E [20–22]. Carotenoids play a role in preventing or delaying degenerative diseases such as cancer and atherosclerosis diseases [23–25], and may be useful in the development of functional foods [15].

There is a paucity of information regarding combining the nutritional compounds of marine-based material and whole-grains. Therefore, the present study is the first to show the glycaemic glucose equivalents (GGE) as a predictor of glycaemic response, antioxidant capacities and physical properties of cereal and *Hematococcus pluvialis* in a model food.

2. Materials and Methods

2.1. Sample Collection and Preparation

Driedmicroalgae *Hematococcus pluvialis* was provided by Supreme Biotechnologies Ltd. (Nelson, New Zealand) and ground using a grinder (AutoGrinder, M-EM0415, Sunbeam Corp Ltd., Auckland, New Zealand). The ground material was sieved through a 0.5 mm screen to obtain flour. Wholemeal wheat (Champion Flour, Auckland, New Zealand), barley (Ceres Organics, Auckland, New Zealand) and oat flours (Ceres Organics, Auckland, New Zealand) were purchased locally.

2.2. Cookie Preparation

Cookies were prepared following the standard American Association of Cereal Chemistry (AACC) method 10–50D [26] with slight modification. Table 1 illustrates the dry ingredients used (sugar, salt and sodium bicarbonate). All dry ingredients (except flour) were mixed in an electric mixer (Breville, Melbourne, Australia) with vegetable shortening (Kremelta, Peerless foods, Braybook, Australia) for 3 min on speed 1. Dextrose solution (8.9 g dextrose anhydrous in 150 mL water) and distilled water were added to the mixer and mixed for a further 1 min on speed 2 with scraping down every 30 s. The flour was added and mixed for 2 min with scraping down every 30 s. The experimental samples were prepared by replacing the wholemeal flour with astaxanthin powder 5%, 10% and 15%. The cookie dough was rolled to a 6 mm thickness using measuring roller and cut with a 57 mm diameter cookie cutter. The cookies were placed on metal trays and baked in a preheated electric oven (BAKBAR turbofan convection oven, E3111, Moffat Pty Ltd., Rolleston, New Zealand) for 8 min at 180 °C. The cookies were cooled at room temperature, placed in air-tight plastic bags and stored at room temperature for 24 h prior to laboratory analysis.

Table 1. Model food formulation.

Sample	Wholemeal Flour (g)	Astaxanthin Powder (g)	Other Ingredients
Control	225.00	-	Vegetable shortening (64.0 g), sugar (130 g), salt (2.1 g), sodium bicarbonate (2.5 g), dextrose solution (33 g), water (16 g)
5% Astaxanthin powder	213.75	11.25	Vegetable shortening (64.0 g), sugar (130 g), salt (2.1 g), sodium bicarbonate (2.5 g), dextrose solution (33 g), water (16 g)
10% Astaxanthin powder	202.50	22.50	
15% Astaxanthin powder	191.25	33.75	

2.3. Physical Characteristics

Cookie diameter (mm) and thickness (mm) were measured using calipers (INSIZE digital caliper, series 1112, INSIZE Inc., Loganville, GA, USA). The colour of the cookie samples were measured in terms of Comission Internationale de l'Eclairage (CIE) L^*, a^* and b^* systems by using a colorimeter (Konica Minolta, Chroma Meters CR-210, Tokyo, Japan). The colour differences of the cookies were calculated by the following equation.

$$\Delta E = \sqrt{(\Delta L^*)^2 + (\Delta a^*)^2 + (\Delta b^*)^2}$$

2.4. Texture

The hardness of the cookies (fracture force) was measured by using a texture analyser (TA.XT plus Texture Analyser, Stable Micro Systems, Godalming, UK) with a 3-point bend rig. The analyser was set at a load cell 50 kg; pre-test speed 2 mm/s; test speed 5 mm/s; post-test speed 10 mm/s; return to start mode. The whole cookies were placed on the support ring and the probe moved downward until the samples were broken. The peak force (kg) was recorded as hardness. Measurements were made in triplicate.

2.5. Moisture

Moisture content of the cookie samples were measured after drying cookie ground samples (2 g) overnight in an oven at 105 °C.

2.6. Determination of Total Phenolic Content

The content of total phenolics of samples was measured by Folin-Ciocalteu reagent (mixture of phosphotungstic and phosphomolybdic acid; that is reduced by phenolics forming a blue complex) using the method described by Floegel et al., 2011 [27] with some modifications. The ground samples (1 g) were dispersed in 20 mL of 70% methanol (by placing on a stirrer overnight). The sample mixture was centrifuged at 700 g Relative centrifugal force (RCF) for 10 min and the supernatant collected to determine the total phenolics. Crude extracts (0.5 mL) were mixed thoroughly with freshly prepared 0.2 N Folin-Ciocalteu's reagent (2.5 mL), followed by 2.0 mL of 7.5% sodium carbonate (Na_2CO_3) and incubated in the dark for 2 hours. The absorbance reaction mixture was measured at 760 nm. Gallic acid (gallic acid, 97%, CAS: 149-91-7, Sigma-Aldrich, St. Louise, MO, USA) was used as a standard and results were expressed as mg gallic acid equivalent (GAE) per g sample.

2.7. Antioxidant Properties

The antioxidant capacity of the samples was measured by the DPPH (2,2-diphenyl-1-picrylhydrazyl) assay as described by Floegel et al., 2011 [27] with some modifications. Briefly, 0.5 mL of crude extract was mixed with freshly prepared 1 mL of 0.1 mM methanolic DPPH (CAS: 1898-66-4, Sigma-Aldrich, St. Louise, MO, USA) solution and incubated in the dark at room temperature for 30 min. The reaction mixture absorbance was measured at 517 nm. In order to calculate the DPPH radical scavenging capacity, trolox (CAS: 53188-07-1, ACROS Organics™, Morris, NJ, USA) was used as a standard and result were expressed as µmol trolox equivalent (TE) per g sample.

Oxygen radical absorbance capacity (ORAC) was determined as described by Floegel et al., 2011 with some modifications. Briefly, 25 µL diluted extract were mixed with 150 µL of 10 nM fluorescein into the microplate well and incubated for 30 min at 37 °C temperature. Twenty five microlitres AAPH (2,2-azobis (2-amidinopropane) dihydrochloride) (CAS: 2997-92-4, Cayman Chemical, Ann Arbor, MI, USA) solution was added to the pre-incubated reaction mixture. Fluorescence was measured (excitation 485 nm; emission 510 nm) from the bottom microplate every 60 s for a total of 60 min. Data analysed by using Omega MARS data analysis software (program version 3.02 R2, BMG Labtech, Mornington, Australia), in order to calculate antioxidant capacity, trolox was used as a standard and results were expressed as a mmol trolox equivalent (TE) per g sample.

2.8. In Vitro Carbohydrate Digestion (Glycaemic Glucose Equivalent-GGE) Analysis

The in vitro digestion process was carried out with the method developed by Foschia, Peressini, Sensidoni, Brennan and Brennan, 2015 [28] and used by Gao, J.R. et al., 2016 [29]. The method estimates the glucose released from the cookie samples during enzymatic hydrolysis over 120 min to predict glycaemic response. In brief: digestions were held in 60 mL plastic pots placed on a controlled temperature stirring hot plate (IKA RT 15, IKA Werke GmbH & Co. KG, Mendelheim, Germany). The samples (0.5 g) were mixed with 30 mL of reverse osmosis water and kept at 37 °C for 10 min with constant stirring on a magnetic starrier. Pepsin solution (1 mL of 1 g pepsin in 10 mL 0.05 M hydrogen chloride (HCl) was added and incubated for 30 min at 37 °C. Aliquots (1 mL) were collected (time 0) from the digestion mixture and added to 4 mL alcohol to arrest enzyme reaction. Amyloglucosidase (0.1 mL) was added to the digestion mixture to prevent end product inhibition of pancreatic α-amylase. Then pancreatin solution (5 mL of 2.5% pancreatin in 0.1 M Malate buffer pH 6.0) was added to the mixture. Further 1 mL aliquots were collected at 20, 60 and 120 min and treated as before, then stored at 4 °C until reducing sugar analysis was carried out. The 3,5-dinitrosalicylic acid (DNS) method was followed to measure reducing sugar content of the samples during in vitro digestion. Glucose release was calculated in mg glucose/g sample and plotted against time and area under the curve (AUC) was calculated by dividing the graph into trapezoids.

2.9. Statistical Analysis

All data was analysed by using the data analysis software, Minitab (version 17, Minitab Inc., State College, PA, USA) to establish significant differences. Analysis of Variance (One-way) was employed with Tukey's test at 95% confidence interval ($p < 0.05$) in all cases. All values were presented as the mean of triplicate determinations $\pm$ standard deviation.

3. Results and Discussion

Cookies were prepared using astaxanthin powder and wholemeal flour. The effects of astaxanthin powder replacement on the physical properties and functional properties of wholemeal flour cookies were analysed.

3.1. Physical Properties of Cookies

The physical characteristics of the cookies are summarised in Table 2. The results showed a significant reduction ($p < 0.05$) in the height and diameter gain of the cookies containing astaxanthin and a significant reduction of ($p < 0.05$) weight loss of the wheat and oat flour cookies with 15% addition of astaxanthin. As the amount of astaxanthin powder increased the weight loss, height and diameter decreased. The largest height changes were observed in cookies made from wholemeal wheat flour, and the largest diameter changes were observed in cookies made from wholemeal oat flour. This observation could be attributed to the hydrophilic nature of the ingredients [30]. The spread factor of a cookie is affected by dough viscosity as well as the acid-base reaction of the ingredients (sodium bicarbonate and fat), causing bubbles in the dough to expand in volume [31]. Physical evaluation of the cookies reported by [32,33], suggested that the spread factor is affected by the water holding capacity of the ingredients. Cookies made with wholemeal barley had increased moisture content with increasing astaxanthin addition. The reason for this phenomenon is that the physical state of starch, protein and fibre are the key determinants of the water holding capacity of the flour as suggested in other papers [34–36]. The moisture content of wheat, oat and barley cookies increased significantly ($p < 0.05$) at all levels of astaxanthin addition (Table 2). This can be attributed to differences in water holding capacity of the ingredients especially different flours [37]. Correspondingly, the hardness of the cookies decreased with the addition of astaxanthin (Table 2). The study indicated that when astaxanthin was incorporated into wheat and oat cookies they were softer and barley cookies were harder in comparison to control cookies. This suggests that water holding capacity of astaxanthin is intermediary between oat and barley flour and it could be due to the nature of the starch and starch-protein interface of different flour. The [38] found that differences in swelling behaviour of the starch granules resulted in cookies with different textural properties, while [34] showed that increased protein content affected the interaction of starch and protein and their hydrogen bonding during dough development.

Table 2. Physical characteristics (after baking: changes in height (%), diameter (%) and weight loss (%); moisture content (%) and hardness (kg) of the model cookies).

Sample	Increase in Height (%)	Increase in Diameter (%)	Weight Loss (%)	Moisture Content (%)	Hardness (kg)
WCC	94.39 ± 3.06 [a]	3.93 ± 0.226	9.71 ± 0.04 [a]	7.50 ± 0.11 [c]	9.26 ± 0.13 [a]
W5A	71.44 ± 8.39 [b]	2.96 ± 1.139 [b]	9.63 ± 0.02 [a,b]	7.83 ± 0.01 [b]	7.79 ± 0.16 [b]
W10A	59.39 ± 3.06 [b,c]	2.27 ± 0.216 [b]	9.48 ± 0.12 [a,b]	7.91 ± 0.07 [b]	7.35 ± 0.58 [b]
W15A	52.94 ± 0.75 [c]	1.15 ± 0.925	9.44 ± 0.12 [b]	8.21 ± 0.03 [a]	7.06 ± 0.48 [b]
BCC	94.33 ± 6.78 [a]	5.23 ± 1.168	10.31 ± 0.11 [a]	7.74 ± 0.02 [d]	4.12 ± 0.12 [c]
B5A	83.50 ± 1.04 [a,b]	4.76 ± 0.444	10.41 ± 0.11 [a]	7.79 ± 0.02 [c]	4.98 ± 0.20 [b]
B10A	74.61 ± 2.91 [b,c]	3.67 ± 0.731 [b]	10.46 ± 0.20 [a]	7.90 ± 0.01 [b]	5.26 ± 0.22 [b]
B15A	65.50 ± 4.84 [c]	2.72 ± 0.314	10.67 ± 0.28 [a]	8.17 ± 0.02 [a]	6.21 ± 0.10 [a]
OCC	70.94 ± 0.91 [a]	23.20 ± 0.25 [a]	11.54 ± 0.17 [a]	5.55 ± 0.05 [d]	7.57 ± 0.05 [a]
O5A	67.94 ± 2.46 [a,b]	12.81 ± 0.49 [a]	11.14 ± 0.23 [a,b]	6.09 ± 0.04 [c]	7.23 ± 0.14 [a,b]
O10A	64.55 ± 0.25 [b]	7.80 ± 0.05 [c]	10.63 ± 0.22 [b,c]	6.66 ± 0.04 [b]	7.02 ± 0.12 [b]
O15A	55.55 ± 0.91 [c]	3.62 ± 0.14 [d]	10.23 ± 0.23 [c]	7.12 ± 0.09 [a]	6.16 ± 0.23 [c]

Data are presented as mean ± standard deviation, n = 3; (a–d): Means within same columns for same flour cookie group that do not share the same superscript are significantly different ($p < 0.05$). W, Wheat; B, Barley; O, Oat; CC, Cookie Control; A, Astaxanthin (5%, 10% or 15%).

3.2. Colour

The colour profile of the cookie samples (surface and ground) are summarised in Table 3. Both the surface colour and the total colour (represented by the ground sample) were measured to determine if there was any interaction in terms of food addition and colour enhancement. The addition of astaxanthin to three types of flour cookies significantly ($p < 0.05$) decreased the lightness (L^*), causing the cookies to became red (a^*) and decreased yellowness (b^*). There was a significant colour change as illustrated by the $\triangle E$ value of the three kind of flour cookies in the following order: control >5% astaxanthin >10% astaxanthin >15% astaxanthin cookies. The main factor causing the colour change of the cookies is due to the pigment of astaxanthin powder, as the level of substitution increased lightness of the cookies decreased and greenness increased. However, the reaction between reducing sugars and amino acids (maillard reaction; starch dextrinization and caramelization) which is induced by heating during baking time also enhances darkness the cookie colour [39] as reflected in colour change (Table 3; $\triangle E$ value).

3.3. Total Phenolic Content (TPC) and Antioxidant Activity of Cookies

The phenolic content, DPPH radical scavenging and ORAC activity of the cookies are summarised in Table 4. It can be seen that the phenolic content increased significantly ($p < 0.05$), and proportionately, with the replacement of astaxanthin powder. This phenomenon is likely to be due to the high amount of phenolic compounds present in astaxanthin (10,000–40,000 mg/kg). Spiller and Dewell (2003) [40] and Sharma and Gujral (2014) [41] have shown that wheat flour has less phenolic compounds compared to barley and oat flour. ORAC values were observed to increase as astaxanthin increased in the formulation (Table 4). Increasing the level of astaxanthin in cookies resulted in a significant increase ($p < 0.05$) of DPPH scavenging activity. These results are due to the addition of astaxanthin derived from microalgae. Previous research has illustrated that astaxanthin compounds are 10 times stronger than the other carotenoids [20] in terms of phenolic antioxidant activities.

Table 3. The CIE colour profiles of the cookies.

Sample	L*	a*	b*	△E
Surface Cookie colour				
WCC	90.40 ± 0.42 [a]	−5.79 ± 0.35 [a]	33.05 ± 0.09 [a]	96.43 ± 0.44 [a]
W5A	84.14 ± 0.26 [b]	−7.37 ± 0.08 [b]	29.16 ± 0.23 [b]	89.36 ± 0.32 [b]
W10A	82.20 ± 0.10 [c]	−8.72 ± 0.20 [c]	27.42 ± 0.06 [c]	87.09 ± 0.13 [c]
W15A	81.42 ± 0.32 [c]	−8.27 ± 0.08 [c]	26.45 ± 0.34 [d]	86.01 ± 0.40 [d]
BCC	94.43 ± 0.45 [a]	−8.16 ± 0.76 [a]	34.55 ± 0.27 [a]	100.89 ± 0.58 [a]
B5A	86.97 ± 0.19 [b]	−9.25 ± 0.04 [b]	32.16 ± 0.07 [b]	93.19 ± 0.20 [b]
B10A	84.77 ± 0.23 [c]	−7.88 ± 0.25 [a]	30.06 ± 0.15 [c]	90.29 ± 0.27 [c]
B15A	83.10 ± 0.14 [d]	−7.83 ± 0.01 [a]	27.95 ± 0.14 [d]	88.02 ± 0.18 [d]
OCC	91.31 ± 0.69 [a]	−5.64 ± 0.34 [a]	35.24 ± 0.15 [a]	98.04 ± 0.71 [a]
O5A	84.19 ± 0.14 [b]	−7.63 ± 0.34 [b]	30.64 ± 0.15 [b]	89.91 ± 0.20 [b]
O10A	82.21 ± 0.13 [c]	−7.87 ± 0.13 [b]	28.26 ± 0.11 [c]	87.31 ± 0.17 [c]
O15A	80.77 ± 0.15 [d]	−8.17 ± 0.13 [b]	26.56 ± 0.19 [d]	85.39 ± 0.22 [d]
Ground Cookie colour				
WCC	87.20 ± 0.20 [a]	−0.32 ± 0.03 [a]	45.33 ± 0.10 [a]	97.55 ± 0.30 [a]
W5A	77.82 ± 0.06 [b]	−6.32 ± 0.04 [c]	43.72 ± 0.26 [b]	90.29 ± 0.01 [b]
W10A	75.42 ± 0.22 [c]	−6.98 ± 0.03 [d]	38.31 ± 0.15 [c]	84.88 ± 0.13 [c]
W15A	69.10 ± 0.30 [d]	−6.20 ± 0.02 [b]	34.13 ± 0.12 [d]	77.32 ± 0.22 [d]
BCC	95.13 ± 0.07 [a]	−13.41 ± 0.21 [c]	43.14 ± 0.41 [a]	105.32 ± 0.13 [a]
B5A	82.96 ± 0.62 [b]	−5.86 ± 0.23 [a]	43.95 ± 0.65 [a]	94.06 ± 0.30 [b]
B10A	74.46 ± 0.63 [c]	−6.56 ± 0.12 [b]	44.02 ± 0.09 [a]	86.74 ± 0.56 [c]
B15A	71.13 ± 0.77 [d]	−6.04 ± 0.38 [a,b]	39.85 ± 1.38 [b]	81.76 ± 0.24 [d]
OCC	93.15 ± 0.59 [a]	−9.32 ± 0.1.34 [b]	49.41 ± 1.35 [a]	105.86 ± 0.02 [a]
O5A	85.92 ± 0.27 [b]	−8.36 ± 0.09 [a,b]	39.58 ± 0.21 [b,c]	94.97 ± 0.16 [b]
O10A	73.24 ± 0.36 [c]	−7.33 ± 0.03 [a]	36.04 ± 0.07 [c]	81.96 ± 0.29 [c]
O15A	71.19 ± 0.47 [d]	−6.99 ± 0.57 [a]	41.34 ± 2.68 [b]	82.64 ± 0.97 [c]

L*, lightness (0 = black, 100 = white); a*, red (+) to green (-); b*, yellow (+) to blue (-); △E, colour difference. Data are presented as mean ± standard deviation, n = 3; (a–d), Means within same columns for same flour cookie group that do not share the same superscript are significantly different ($p < 0.05$). W, Wheat; B, Barley; O, Oat; CC, Cookie Control; A, Astaxanthin (5%, 10% or 15%).

Table 4. Total phenolic content and antioxidant capacity.

Sample	TPC (mg GAE/g Sample)	DPPH (µmol TE/g Sample)	ORAC (mmol TE/g Sample)
WCC	0.59 ± 0.01 [d]	0.54 ± 0.01 [d]	0.09 ± 0.001 [b]
W5A	0.80 ± 0.01 [c]	0.95 ± 0.03 [c]	0.11 ± 0.001 [a]
W10A	0.95 ± 0.01 [b]	1.10 ± 0.01 [b]	0.12 ± 0.001 [a]
W15A	1.14 ± 0.01 [a]	1.26 ± 0.03 [a]	0.12 ± 0.004 [a]
BCC	0.63 ± 0.01 [c]	1.36 ± 0.01 [d]	0.08 ± 0.003 [b]
B5A	0.95 ± 0.02 [b]	1.69 ± 0.02 [c]	0.09 ± 0.002 [a]
B10A	1.15 ± 0.09 [a]	1.74 ± 0.01 [b]	0.09 ± 0.002 [a]
B15A	1.27 ± 0.01 [a]	1.79 ± 0.01 [a]	0.10 ± 0.002 [a]
OCC	0.87 ± 0.01 [d]	1.13 ± 0.01 [d]	0.08 ± 0.001 [c]
O5A	1.03 ± 0.01 [c]	1.22 ± 0.01 [c]	0.10 ± 0.002 [b]
O10A	1.28 ± 0.01 [b]	1.34 ± 0.01 [b]	0.10 ± 0.001 [a]
O15A	1.44 ± 0.01 [a]	1.46 ± 0.01 [a]	0.11 ± 0.001 [a]

Data are presented as mean ± standard deviation, n = 3; (a–d), Means within same columns for same flour cookie group do not share the same superscript are significantly different ($p < 0.05$). W, Wheat; B, Barley; O, Oat; CC, Cookie Control; A, Astaxanthin (5%, 10% or 15%).

3.4. Glycaemic Glucose Equivalent (GGE) Analysis

Figure 1 illustrates the in vitro digestion of cookies, calculated as the amount of reducing sugar released by digestive enzymes over 120 min. All the samples demonstrated the impact of the substitution of astaxanthin in the following order (5% > 10% > 15%) and significantly slowed the amount of reducing sugar released (calculated as mg glucose/g sample of incremental area under the curve (iAUC)) as compare with the control cookies.

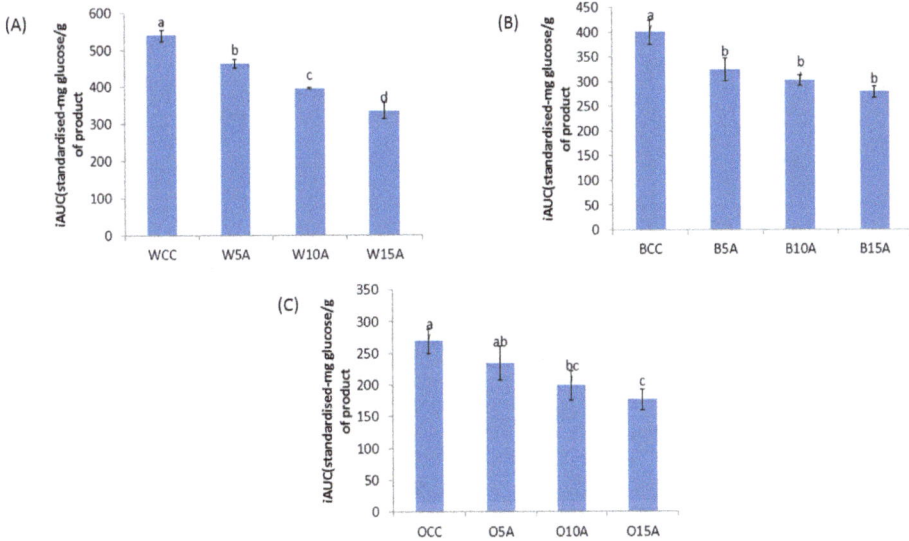

Figure 1. Reducing sugar released (mg/g sample) after 120 min digestion of (**A**) wheat, (**B**) barley and (**C**) oat wholemeal flour cookies with astaxanthin substitution. WCC, wheat cookie control; W5A, wheat + 5% astaxanthin cookie; W10A, wheat + 10% astaxanthin cookie; W15A, wheat + 5% astaxanthin cookie; BCC, barley cookie control; B5A, barley + 5% astaxanthin cookie; B10A, barley + 10% astaxanthin cookie; B15A, barley + 15% astaxanthin cookie; OCC, oat cookie control; O5A, oat + 5% astaxanthin cookie; O10A, oat + 10% astaxanthin cookie; O15A, oat + 15% astaxanthin cookie. (a–d), Means within same figure that do not share the same superscript are significantly different ($p < 0.05$).

It is possible that the high antioxidant activity of the astaxanthin powder could be related to the decreased rate of sugar released [21]. Researchers have shown that antioxidants can impair enzyme activity during the digestion [42]. The interaction between phenolic compounds and digestive enzymes [43] could affect the non-covalent starch-phenolic interactions thus impeding starch degradation [44,45]. Additionally, the rate of sugar release may also be decreased due to the non-starchy network of fibre and protein in the system which entraps starch granules and acts as a physical barrier thus limiting enzyme accessibility [28,46].

Figure 2 illustrates the rate of reaction of starch conversion to reducing sugar release over the 120 min in vitro digestion period. Form this figure it can be observed that the rate of reaction between 20–120 min appears to be greater for the control samples as compared with the samples containing astaxanthin. It can also be observed that the oat samples generally showed a lower sugar release profile than the barley and the wheat samples. It is possible that the in vitro digestion studies observed in Figures 1 and 2 are related to the total phenolic content/antioxidant activity of the samples (Table 4). Further work is required to determine whether this is an indirect relationship or if there is a mechanistic association between phenolic content of the cookies and reduced starch digestion.

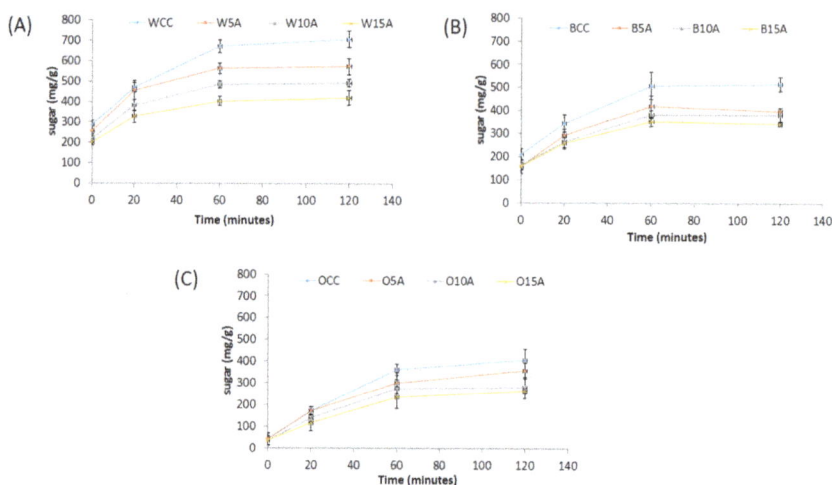

Figure 2. Reducing sugar released (mg/g sample) during the 120 min in vitro digestion process of (**A**) wheat, (**B**) barley and (**C**) oat wholemeal flour cookies with astaxanthin substitution. WCC, wheat cookie control; W5A, wheat + 5% astaxanthin cookie; W10A, wheat + 10% astaxanthin cookie; W15A, wheat + 5% astaxanthin cookie; BCC, barley cookie control; B5A, barley + 5% astaxanthin cookie; B10A, barley + 10% astaxanthin cookie; B15A, barley + 15% astaxanthin cookie; OCC, oat cookie control; O5A, oat + 5% astaxanthin cookie; O10A, oat + 10% astaxanthin cookie; O15A, oat + 15% astaxanthin cookie.

4. Conclusions

The research has illustrated the possible use of novel natural ingredients in alerting the functional quality and biological activity of simple foods. In particular, in vitro digestion (GGE analysis) of the cookies demonstrated significantly lower glucose release when astaxanthin increased in the formulation. The results also demonstrated that the combination of astaxanthin with wholemeal flour significantly improve the antioxidant properties of the cookies. Thus, the inclusion of astaxanthin illustrates a potential synergy between microalgae and wholemeal flour of the model food. As such this combination can contribute to the intake of natural bioactive compounds in the human diets for the potential health benefits.

Acknowledgments: The supply of microalgae by Supreme Biotechnologies Ltd., Nelson, New Zealand is kindly acknowledged. This research was funded by Lincoln University, New Zealand and National Science Challenge High Value Nutrition Award.

Author Contributions: A.K.M.M.H. and M.A.B. conducted the experiments. All authors were involved in the experimental design, data analysis, drafting, reading and approving the final manuscript.

Conflicts of Interest: The authors declare no conflict of interest.

References

1. Andersson, A.A.M.; Dimberg, L.; Aman, P.; Landberg, R. Recent findings on certain bioactive components in whole grain wheat and rye. *J. Cereal Sci.* **2014**, *59*, 294–311. [CrossRef]
2. Brennan, C.S.; Cleary, L.J. The potential use of cereal (1→3,1→4)-β-D-glucans as functional food ingredients. *J. Cereal Sci.* **2005**, *42*, 1–13. [CrossRef]
3. Brennan, C.S. Dietary fibre, glycaemic response, and diabetes. *Mol. Nutr. Food Res.* **2005**, *49*, 560–570. [CrossRef]
4. Ye, E.Q.; Chacko, S.A.; Chou, E.L.; Kugizaki, M.; Liu, S.M. Greater Whole-grain intake is associated with lower risk of type 2 diabetes, cardiovascular disease, and weight gain. *J. Nutr.* **2012**, *142*, 1304–1313. [CrossRef] [PubMed]

5. Knudsen, M.D.; Kyro, C.; Olsen, A.; Dragsted, L.O.; Skeie, G.; Lund, E.; Aringman, P.; Nilsson, L.M.; Bueno-de-Mesquita, H.B.; Tjonneland, A.; et al. Self-reported whole-grain intake and plasma alkylresorcinol concentrations in combination in relation to the incidence of colorectal cancer. *Am. J. Epidemiol.* **2014**, *179*, 1188–1196. [CrossRef] [PubMed]

6. Cho, S.S.; Qi, L.; Fahey, G.C.; Klurfeld, D.M. Consumption of cereal fiber, mixtures of whole grains and bran, and whole grains and risk reduction in type 2 diabetes, obesity, and cardiovascular disease. *Am. J. Clin. Nutr.* **2013**, *98*, 594–619. [CrossRef] [PubMed]

7. Zhou, A.L.; Hergert, N.; Rompato, G.; Lefevre, M. Whole grain oats improve insulin sensitivity and plasma cholesterol profile and modify gut microbiota composition in C57BL/6J mice. *J. Nutr.* **2015**, *145*, 222–230. [CrossRef] [PubMed]

8. Pasqualone, A.; Laddomada, B.; Centomani, I.; Paradiso, V.M.; Minervini, D.; Caponio, F.; Summo, C. Bread making aptitude of mixtures of re-milled semolina and selected durum wheat milling by-products. *LWT Food Sci. Technol.* **2017**, *78*, 151–159. [CrossRef]

9. Oliveira, L.C.; Rosell, C.M.; Steel, C.J. Effect of the addition of whole-grain wheat flour and of extrusion process parameters on dietary fibre content, starch transformation and mechanical properties of a ready-to-eat breakfast cereal. *Int. J. Food Sci. Technol.* **2015**, *50*, 1504–1514. [CrossRef]

10. Robin, F.; Theoduloz, C.; Srichuwong, S. Properties of extruded whole grain cereals and pseudocereals flours. *Int. J. Food Sci. Technol.* **2015**, *50*, 2152–2159. [CrossRef]

11. Lu, X.K.; Brennan, M.A.; Serventi, L.; Mason, S.; Brennan, C.S. How the inclusion of mushroom powder can affect the physicochemical characteristics of pasta. *Int. J. Food Sci. Technol.* **2016**, *51*, 2433–2439. [CrossRef]

12. Sobota, A.; Rzedzicki, Z.; Zarzycki, P.; Kuzawinska, E. Application of common wheat bran for the industrial production of high-fibre pasta. *Int. J. Food Sci. Technol.* **2015**, *50*, 111–119. [CrossRef]

13. Grigor, J.M.; Brennan, C.S.; Hutchings, S.C.; Rowlands, D.S. The sensory acceptance of fibre-enriched cereal foods: A meta-analysis. *Int. J. Food Sci. Technol.* **2016**, *51*, 3–13. [CrossRef]

14. Alves, C.; Pinteus, S.; Simoes, T.; Horta, A.; Silva, J.; Tecelao, C.; Pedrosa, R. *Bifurcaria bifurcata*: A key macro-alga as a source of bioactive compounds and functional ingredients. *Int. J. Food Sci. Technol.* **2016**, *51*, 1638–1646. [CrossRef]

15. Kadam, S.U.; Tiwari, B.K.; O'Donnell, C.P. Extraction, structure and biofunctional activities of laminarin from brown algae. *Int. J. Food Sci. Technol.* **2015**, *50*, 24–31. [CrossRef]

16. Aguilera-Morales, M.; Casas-Valdez, M.; Carrillo-Dominguez, B.; Gonzalez-Acosta, B.; Perez-Gil, F. Chemical composition and microbiological assays of marine algae *Enteromorpha* spp. as a potential food source. *J. Food Compost. Anal.* **2005**, *18*, 79–88. [CrossRef]

17. Lordan, S.; Ross, R.P.; Stanton, C. Marine bioactives as functional food ingredients: Potential to reduce the incidence of chronic diseases. *Mar. Drugs* **2011**, *9*, 1056–1100. [CrossRef] [PubMed]

18. Plaza, M.; Cifuentes, A.; Ibanez, E. In the search of new functional food ingredients from algae. *Trends Food Sci. Technol.* **2008**, *19*, 31–39. [CrossRef]

19. Wu, H.Y.; Hong, H.L.; Zhu, N.; Han, L.M.; Suo, Q.L. Two ethoxyquinoline metabolites from the alga *Heamatococcus pluvialis*. *Chem. Nat. Compd.* **2014**, *50*, 578–580. [CrossRef]

20. Miki, W. Biological functions and activities of animal carotenoids. *Pure Appl. Chem.* **1991**, *63*, 141–146. [CrossRef]

21. Naguib, Y.M.A. Antioxidant activities of astaxanthin and related carotenoids. *J. Agric. Food Chem.* **2000**, *48*, 1150–1154. [CrossRef] [PubMed]

22. Shimidzu, N.; Goto, M.; Miki, W. Carotenoids as singlet oxygen quenchers in marine organisms. *Fish. Sci.* **1996**, *62*, 134–137.

23. Barros, M.P.; Poppe, S.C.; Bondan, E.F. Neuroprotective properties of the marine carotenoid astaxanthin and omega-3 fatty acids, and perspectives for the natural combination of both in krill oil. *Nutrients* **2014**, *6*, 1293–1317. [CrossRef] [PubMed]

24. Raposo, M.F.D.; de Morais, A.; de Morais, R. Carotenoids from marine microalgae: A valuable natural source for the prevention of chronic diseases. *Mar. Drugs* **2015**, *13*, 5128–5155. [CrossRef] [PubMed]

25. Riccioni, G.; D'Orazio, N.; Franceschelli, S.; Speranza, L. Marine carotenoids and cardiovascular risk markers. *Mar. Drugs* **2011**, *9*, 1166–1175. [CrossRef] [PubMed]

26. American Association of Cereal Chemists. *Approved Methods of the American Association of Cereal Chemists*, 9th ed.; American Association of Cereal Chemists: St. Paul, MN, USA, 1995.

27. Floegel, A.; Kim, D.O.; Chung, S.J.; Koo, S.I.; Chun, O.K. Comparison of ABTS/DPPH assays to measure antioxidant capacity in popular antioxidant-rich US foods. *J. Food Compost. Anal.* **2011**, *24*, 1043–1048. [CrossRef]

28. Foschia, M.; Peressini, D.; Sensidoni, A.; Brennan, M.A.; Brennan, C.S. Synergistic effect of different dietary fibres in pasta on in vitro starch digestion? *Food Chem.* **2015**, *172*, 245–250. [CrossRef] [PubMed]

29. Gao, J.R.; Brennan, M.A.; Mason, S.L.; Brennan, C.S. Effect of sugar replacement with stevianna and inulin on the texture and predictive glycaemic response of muffins. *Int. J. Food Sci. Technol.* **2016**, *51*, 1979–1987. [CrossRef]

30. Okpala, L.; Okoli, E.; Udensi, E. Physico-chemical and sensory properties of cookies made from blends of germinated pigeon pea, fermented sorghum, and cocoyam flours. *Food Sci. Nutr.* **2013**, *1*, 8–14. [CrossRef] [PubMed]

31. Chung, H.J.; Cho, A.; Lim, S.T. Utilization of germinated and heat-moisture treated brown rices in sugar-snap cookies. *LWT Food Sci. Technol.* **2014**, *57*, 260–266. [CrossRef]

32. Brennan, C.S.; Samyue, E. Evaluation of starch degradation and textural characteristics of dietary fiber enriched biscuits. *Int. J. Food Prop.* **2004**, *7*, 647–657. [CrossRef]

33. Giami, S.Y.; Achinewhu, S.C.; Ibaakee, C. The quality and sensory attributes of cookies supplemented with fluted pumpkin (*Telfairia occidentalis* Hook) seed flour. *Int. J. Food Sci. Technol.* **2005**, *40*, 613–620. [CrossRef]

34. Mais, A.; Brennan, C.S. Characterisation of flour, starch and fibre obtained from sweet potato (kumara) tubers, and their utilisation in biscuit production. *Int. J. Food Sci. Technol.* **2008**, *43*, 373–379. [CrossRef]

35. Ragaee, S.; Abdel-Aal, E.S.M. Pasting properties of starch and protein in selected cereals and quality of their food products. *Food Chem.* **2006**, *95*, 9–18. [CrossRef]

36. Yamsaengsung, R.; Berghofer, E.; Schoenlechner, R. Physical properties and sensory acceptability of cookies made from chickpea addition to white wheat or whole wheat flour compared to gluten-free amaranth or buckwheat flour. *Int. J. Food Sci. Technol.* **2012**, *47*, 2221–2227. [CrossRef]

37. Inglett, G.E.; Chen, D.J.; Liu, S.X. Physical properties of gluten-free sugar cookies made from amaranth-oat composites. *LWT Food Sci. Technol.* **2015**, *63*, 214–220. [CrossRef]

38. Kweon, M.; Slade, L.; Levine, H. Solvent retention capacity (SRC) testing of wheat flour: Principles and value in predicting flour functionality in different wheat-based food processes and in wheat breeding—A review. *Cereal Chem.* **2011**, *88*, 537–552. [CrossRef]

39. Chevallier, S.; Colonna, P.; Lourdin, D. Contribution of major ingredients during baking of biscuit dough systems. *J. Cereal Sci.* **2000**, *31*, 241–252. [CrossRef]

40. Spiller, G.A.; Dewell, A. Safety of an astaxanthin-rich *Haematococcus pluvialis* algal extract: A randomized clinical trial. *J. Med. Food* **2003**, *6*, 51–56. [CrossRef] [PubMed]

41. Sharma, P.; Gujral, H.S. Cookie making behavior of wheat-barley flour blends and effects on antioxidant properties. *LWT Food Sci. Technol.* **2014**, *55*, 301–307. [CrossRef]

42. Matsiu, T.; Ebuchi, S.; Kobayashi, M.; Fukui, K.; Sugita, K.; Terahara, N.; Matsumoto, K. Anti-hyperglycemic effect of diacylated anthocyanin derived from Ipomoea batatas cultivar Ayamurasaki can be achieved through the alpha-glucosidase inhibitory action. *J. Agric. Food Chem.* **2002**, *50*, 7244–7248. [CrossRef]

43. Paliwal, C.; Ghosh, T.; Bhayani, K.; Maurya, R.; Mishra, S. Antioxidant, anti-nephrolithe activities and in vitro digestibility studies of three different cyanobacterial pigment extracts. *Mar. Drugs* **2015**, *13*, 5384–5401. [CrossRef] [PubMed]

44. Bordenave, N.; Hamaker, B.R.; Ferruzzi, M.G. Nature and consequences of non-covalent interactions between flavonoids and macronutrients in foods. *Food Funct.* **2014**, *5*, 18–34. [CrossRef] [PubMed]

45. Soong, Y.Y.; Tan, S.P.; Leong, L.P.; Henry, J.K. Total antioxidant capacity and starch digestibility of muffins baked with rice, wheat, oat, corn and barley flour. *Food Chem.* **2014**, *164*, 462–469. [CrossRef] [PubMed]

46. Wolter, A.; Hager, A.S.; Zannini, E.; Arendt, E.K. Influence of sourdough on in vitro starch digestibility and predicted glycemic indices of gluten-free breads. *Food Funct.* **2014**, *5*, 564–572. [CrossRef] [PubMed]

foods

MDPI

Article

Sodium Chloride and Its Influence on the Aroma Profile of Yeasted Bread

Markus C. E. Belz [1], Claudia Axel [1], Jonathan Beauchamp [2], Emanuele Zannini [1], Elke K. Arendt [1,*] and Michael Czerny [2]

[1] School of Food and Nutritional Sciences, University College Cork, National University of Ireland, College Road, T12 Y337 Cork, Ireland; belz.markus@web.de (M.C.E.B.); c.axel@umail.ucc.ie (C.A.); e.zannini@ucc.ie (E.Z.)

[2] Department of Sensory Analytics, Fraunhofer Institute for Process Engineering and Packaging IVV, 85354 Freising, Germany; jonathan.beauchamp@ivv.fraunhofer.de (J.B.); michael.czerny@ivv.fraunhofer.de (M.C.)

* Correspondence: e.arendt@ucc.ie; Tel.: +353-21-490-2064; Fax: +353-21-427-0213

Academic Editor: Anthony Fardet
Received: 30 June 2017; Accepted: 9 August 2017; Published: 12 August 2017

Abstract: The impact of sodium chloride (NaCl) concentration on the yeast activity in bread dough and its influence on the aroma profile of the baked bread was investigated. Key aroma compounds in the bread samples were analysed by two-dimensional high-resolution gas chromatography-mass spectrometry in combination with solvent-assisted flavour evaporation distillation. High-sensitivity proton-transfer-reaction mass spectrometry was used to detect and quantify 2-phenylethanol in the headspace of the bread dough during fermentation. The analyses revealed significant ($p < 0.05$) changes in the aroma compounds 2-phenylethanol, (E)-2-nonenal, and 2,4-(E,E)-decadienal. Descriptive sensory analysis and discriminating triangle tests revealed that significant differences were only determinable in samples with different yeast levels but not samples with different NaCl concentrations. This indicates that a reduction in NaCl does not significantly influence the aroma profile of yeasted bread at levels above the odour thresholds of the relevant compounds, thus consumers in general cannot detect an altered odour profile of low-salt bread crumb.

Keywords: descriptive sensory; PTR-MS; GC-MS; Ehrlich pathway; bread aroma; salt reduction chemical compounds: 2-phenylethanol (PubChem CID: 6054); (E)-2-nonenal (PubChem CID: 5283335); 2,4-(E,E)-decadienal (PubChem CID: 5283349)

---

1. Introduction

Sodium chloride (NaCl), or salt, is a major taste contributor to food. A reduction of salt in food products generally leads to less intense taste and flavour. The impact of salt reduction on taste profiles has been demonstrated for numerous foods, amongst them white yeasted bread. An early investigation on white yeasted, rye and rye-sourdough breads indicated a low consumer preference for the reduced-salt breads [1]. In contrast, however, Wyatt [2] observed no significant difference in consumer preference for white bread containing 50% less NaCl than a reference bread. The current challenge for food producers is to develop products with a reduced salt content but an unimpaired and consistent taste. This has been investigated with the use of salt replacers such as potassium chloride, magnesium chloride, ammonium chloride, calcium chloride and calcium carbonate [3,4], the use of sourdough [5], the inclusion of flavour-enhancing acids and other potent aroma compounds [6,7], or by changes to the bread crumb texture that influence the saltiness perception [8–10]. In contrast to taste, less is known about the influence of salt reduction on the volatile aroma profile of food. Volatile aroma compounds impart flavour to food, and the volatile fraction of bread is highly complex with about 600 volatile compounds reported to be present in bread crumb [11]. In particular, the yeast metabolism plays a key role in the development of the aroma profile of bread, and salt, primarily its sodium ions, have a direct impact on yeast activity. In addition to ethanol and carbon dioxide, many low-molecular-weight flavour compounds such as further alcohols, aldehydes, acids, esters, sulphides and carbonyl compounds are produced by the yeast metabolism. These volatile compounds are essential contributors to the flavour of fermented foods and beverages [12,13]. The Ehrlich pathway is one of several routes responsible for the generation of aroma compounds by yeast in bread. In particular it leads to the formation of potent compounds such as fusel alcohols and acids. The efficacy of the Ehrlich pathway in converting amino acids into alcoholic odorants was investigated, amongst others, by Czerny and Schieberle [14], who used stable isotope dilution assays (SIDAs) to demonstrate the conversion of ^{13}C (6)-leucine to the metabolite 3-methylbutanol.

The reducing activity of yeast during bread dough fermentation also has a critical impact on bread aroma, as has been similarly observed during beer wort fermentation [15]. Unsaturated aldehydes such as (E)-2-nonenal and 2,4-(E,E)-decadienal are derived from the oxidation of linoleic acid and are well-known for their contribution to fatty odours in wheat bread [11]. The reduction of these unsaturated volatile compounds by yeast to their corresponding alcohols has an impact on bread aroma; as such, a variation in yeast activity results in an altered aroma of the bread. Notably, it has been observed that *Saccharomyces cerevisiae* fully reduces unsaturated aldehydes to the corresponding alcohols [16].

The present work aimed at determining the impact of NaCl on the yeast metabolism in bread dough during fermentation and its influence on the overall aroma of the bread. More specifically, the unsaturated aldehydes (E)-2-nonenal (fatty) and 2,4-(E,E)-decadienal (fatty), and the alcoholic compound 2-phenylethanol (rose-like) were investigated as the key aroma compounds in bread based on preliminary analyses and supported by reports in the literature [11,17,18]. Recently Martins et al., 2015 evaluated the impact of bread fortification with dry spent yeast from brewing industry on physical, chemical and sensorial characteristics of home-made bread with the goal of increasing its β-glucan content. The sensory analysis showed how only the key odour hexanal was presented a significant increase in fortified bread [19]. A complementary analytical approach using two-dimensional high-resolution gas chromatography-mass spectrometry (2D-HRGC-MS) and proton-transfer-reaction mass spectrometry (PTR-MS) was used to quantify and monitor the generation of the selected aroma compounds in bread crumb samples containing different levels of NaCl and yeast. Sensory analysis was performed on the bread crumb samples to determine their odour characteristics and assess the impact of salt reduction on bread crumb based on a discrimination triangle test.

2. Materials and Methods

2.1. Microbiology

Instant active dry yeast consisting of living cells of *Saccharomyces cerevisiae* (Panté; Puratos, Belgium) was diluted in Ringer solution at a concentration of 10^{-5} g mL^{-1}. An aliquot of 10 μL of the yeast solution was grown as a centre colony on yeast-selective potato dextrose agar plates (Fluka Chemie AG, Buchs, Switzerland) containing different amounts of sodium chloride (NaCl) (0, 0.3%, 1.2%, 2.0%, 3.0%, and 4.0% w/w) at 30 °C for 8 days. The growth rate was recorded every day by measuring the diameter of the visible colonies with an electronic calliper.

2.2. Baking Procedure and Loaf Analyses

Wheat bread was prepared by mixing (spiral mixer, Kenwood KM020, Kenwood Manufacturing Co., Ltd., Hampshire, UK) Baker's flour (Odlums, Dublin, Ireland), yeast, NaCl (Glacia British Salt Limited, Middlewich, UK) at levels of 0, 0.26%, 1.04%, 1.73%, 2.60% and 3.46% (w/w) and tap water (water levels set to 500 Brabender units, BU, depending on the amount of NaCl by using a farinograph, Brabender OHG, Duisburg, Germany). Considering an average bake loss of 13.5% the

NaCl concentrations in the final bread loaves resulted in 0, 0.3%, 1.2%, 2.0%, 3.0% and 4.0% (Table 1). For each of the 10 different batches 6 loaves were prepared. For dough samples with varying amounts of yeast (1.5%, 0.9%, 0.6%, 0.3%), the concentrations of water and NaCl relative to the mass of flour were kept constant at 61.75% and 0.26% (w/w), respectively (Table 1). After bulk fermentation for 15 min at 30 °C and 85% relative humidity, bread loaves (450 ± 1 g) were formed using a molding machine (Machinefabriek Holtkamp B.V., Almelo, The Netherlands). The loaves were then placed into non-stick baking tins (180 mm × 120 mm × 60 mm), fermented for 75 min at 30 °C and 85% relative humidity, and then baked for 35 min at 230 °C (top and bottom heating). The ovens were pre-steamed (0.3 L water) and then steamed when loaded (0.7 L water). After baking, the loaves were removed from the tins and left to cool on cooling racks for 120 min at room temperature. Bake loss and specific volume were measured for all of the baked loaves. The bake loss was determined as the difference in mass between the dough and baked loaf. The specific volume was determined by a 3D laser scan using a VolScan Profiler 300 (Stable Micro Systems, Godalming, UK).

Table 1. Ingredient quantities in the breads containing varying amounts of NaCl at constant yeast (1.2% w/w) and varying amounts of yeast at constant NaCl (0.3% w/w) *.

Ingredient	Ingredient Quantity for Each Type of Bread (% w/w flour)									
	Variable NaCl (% w/w) at 1.2% w/w Yeast						Variable Yeast (% w/w) at 0.3% w/w NaCl			
	0	0.26	1.04	1.73	2.60	3.46	1.5	0.9	0.6	0.3
Wheat flour	100.00	100.00	100.00	100.00	100.00	100.00	100.00	100.00	100.00	100.00
Yeast	2.00	2.00	2.00	2.00	2.00	2.00	2.47	1.48	0.98	0.48
Tap water (30°)	62.95	61.75	61.00	60.90	59.75	59.75	61.75	61.75	61.75	61.75
NaCl	0.00	0.43	1.72	2.87	4.36	5.80	0.43	0.43	0.43	0.43

* The 0.3% NaCl concentration was used in accordance to the Directive 80/777/EEC to claim a bread low in salt.

2.3. Rheofermentometer

A rheofermentometer (RheoF3, Chopin Technologies, Villeneuve-la-Garenne, Paris, France) was used to evaluate carbon dioxide (CO_2) release and dough development of the different dough samples. Samples of 300 g of each dough were prepared in the same manner as described above for baking trials. The tests were performed at 30 °C over a 90 min period. As common practice for wheat dough, a cylindrical weight of 1.5 kg was applied onto the fermentation chamber. The total volume of CO_2, the volume of retention for each sample, the lost volume of CO_2, and the retention coefficient (capability of a dough to retain gas) were determined. Results are presented as the average of triplicate measurements.

2.4. Extraction of Volatile Aroma Compounds

Samples were prepared by cutting the bread crumb into 1 cm^3 cubes, freezing these in liquid nitrogen, and then grinding them using a standard blender. Isotope-labelled standard solutions ([^{2}H$_2$]-(E)-2-nonenal 0.24 µg/mL; [^{2}H$_2$]-2,4-(E,E)-decadienal 0.50 µg/mL; [^{2}H$_{4-5}$]-2-phenylethanol 10.30 µg/mL) were added as internal standard to 50 ± 1 g of the ground crumb and the aroma compounds were extracted with 150 mL dichloromethane that was stirred at 120 rpm for 60 min at room temperature and then filtered to remove the suspension. This extraction step was repeated twice for each 50-g crumb sample and the filtrates were combined. The extracts were purified using solvent-assisted flavour evaporation (SAFE) distillation [20]. The distillates were concentrated down to a volume of 0.1 mL and these were subsequently stored at −20 °C prior to the analysis.

2.5. Two-Dimensional High-Resolution Gas Chromatography-Mass Spectrometry (2D-HRGC-MS)

Quantification of the selected aroma compounds was made using a two-dimensional high-resolution gas chromatography-mass spectrometer (2D-HRGC-MS), with a cryogenic trapping system (CryoTrap; Gerstel, Stadt, Germany) connecting the first GC system (Type 3800, Varian, Darmstadt, Germany) with a preparative DB-5 column to the second GC with a DB-FFAP column (each 30 m × 0.32 mm, 0.25 μm film thickness). The helium carrier gas flow was set to 1.5 mL min^{-1}. The initial temperature of the first GC oven was 40 °C, which was subsequently heated at a rate of 8 °C min^{-1} to 230 °C. The eluting aroma compounds were transferred at defined retention times onto the cryo-trap, which was cooled to −100 °C. The volatiles were then flushed onto the column in the second GC oven by thermal desorption at 250 °C. The temperature of this second oven was increased from 40 °C to 250 °C at a rate of 6 °C min^{-1} and then held for 5 min at 250 °C. The eluting compounds were analysed with a Saturn 2200 mass spectrometer (Varian, Darmstadt, Germany) by chemical ionisation (CI) using methanol as the reagent gas.

2.6. Proton-Transfer-Reaction Mass Spectrometry (PTR-MS)

A high sensitivity proton-transfer-reaction mass spectrometer (hs-PTR-MS; IONICON Analytik GmbH, Innsbruck, Austria) was used to analyse the release of selected aroma compounds from the breads during fermentation. The instrument was operated at an electric field to buffer gas number density ratio (E/N) of 132 Td, which was established with drift tube settings of 600 V, 2.2 mbar and 60 °C. The PTR-MS was set to measure in mass scan mode in the range m/z 20–130 at a dwell time of 500 ms per m/z. Five individual scans of 51 s duration were made per sampling period, resulting in a complete analysis time of 255 s. A 1-m long, 1/16″ OD, 0.04″ ID Silcosteel™ (Restek GmbH, Bad Homburg, Germany) sample inlet line, heated to 65 °C and with a flow of 500 mL min^{-1}, was used to transfer the sample gas to the instrument reaction chamber.

Dough samples of 300 g were placed in 1-L perfluoroalkoxy (PFA) containers (AHF Analysentechnik GmbH, Darmstadt, Germany) for the on-line measurement of volatiles in the headspace of the dough during fermentation at 30 °C and 85% RH over 75 min. Five scan cycles of zero-air—i.e., air free of volatile organic compounds (VOCs)—in the empty sample container were made at the beginning of each analysis to determine the background noise and the limit of detection of the system. The mean signals from these scans were subtracted from the sample signals to correct for this background.

The intensities of the m/z relating to the abundance of the selected aroma compounds in the headspace gas of the sample chamber were converted to approximate concentrations (with an estimated accuracy of ± 30%) using a standard reaction rate (k) of 2.0 × 10^{-9} cm^3/s [21]. The data were screened for m/z 105 specific to 2-phenylethanol, which is ionised to a cation by dehydroxylation and protonation. For the unsaturated aldehydes, the respective molecular ions at m/z 141 for (E)-2-nonenal and m/z 153 for 2,4-(E,E)-decadienal were outside the m/z scan range and thus could not be detected. It might be noted that a certain degree of fragmentation of these two compounds is expected to occur, thus their potential detection at m/z values within the scanned range might have been achievable, but this was hindered by a lack of knowledge of the exact fragments or their potential overlap with other compounds, and instrumental problems that impaired the sensitivity to m/z in the higher range. Thus only 2-phenylethanol will be reported for the PTR-MS data here.

2.7. Sensory Analysis

Sensory analyses of the samples were performed via the aroma profile analysis (APA) technique. Descriptive analyses were performed using a trained panel of 15 members, with at least ten assessors participating in each individual sensory session. The panellists were trained in weekly sessions to recognise the selected aroma compounds according to their odour qualities by smelling reference aqueous aroma solutions at different odorant concentrations. Training was performed over a period of at least six months prior to participation in the actual sensory experiments and the performance of each panellist was assessed via standard procedures.

Bread loaves were cut into 2 cm-thick slices and the crust was removed. The yeasted dough and wheat flour bread samples were presented to the sensory panel for orthonasal assessment after storage in a closed glass beaker for 30 min at room temperature. The perceived odour qualities of the bread crumbs were described as being yeast/dough-like and flour-like based on a comparison to aqueous reference solutions. The panel agreed on the characteristic odour attributes of each sample in a group discussion. The pure compounds used for the reference solutions were purchased from Sigma-Aldrich (Taufkirchen, Germany), Acros (Geel, Belgium) and AromaLab (Freising, Germany). Crumb samples were then presented again to the panel in a second sensory session to evaluate the intensities of the aforementioned odour attributes on a scale from 0 (not detectable) over 1 (weak intensity), 2 (medium intensity) to 3 (high intensity). The sensory score of each attribute was calculated as an arithmetic mean. The assessors were trained immediately prior to the analysis with aqueous odorant solutions at defined super-threshold concentrations (factor 100 above the odour threshold) [18,22].

Sensory triangle tests were additionally performed on selected sample pairs by the panel to determine whether potential odour changes are detectable by consumers. The sample pairs, namely 0.3% and 1.5% yeast, 0.3% and 1.2% NaCl, and 1.2% and 3.0% NaCl were chosen based on a 'standard-salt' level of 1.2% w/w NaCl [23], a "low-salt" level of 0.3% w/w NaCl, and an "high-salt" level of 3.0% w/w NaCl. The bread loaves were sliced and uniform round pieces were punched-out for presentation. The panellists were required to smell the samples and identify which of the three samples differed (forced-choice test). The tests were repeated for all combinations of each sample pair during each of the three independent sessions.

2.8. Statistical Analysis

Statistical analyses were performed on the sensory assessment results using Minitab for Windows statistical analysis software package (Systat Software, Inc., Chicago, IL, USA). The data were subjected to one-way analysis of variance (ANOVA). A Fisher's least significant difference (LSD) test was performed for multiple comparisons for cases when an F-test showed significant differences ($p < 0.05$). Mean values of the three separate experiments with three independent samples from each batch were then calculated and are presented here, unless otherwise stated. The significance level was set to $p < 0.05$ throughout, unless otherwise stated.

3. Results and Discussion

3.1. Microbiology and Rheofermentometer

Initial assessments of the impact of NaCl on the activity and proliferation of the yeast showed an exponential inhibition of yeast proliferation with increasing amounts of NaCl. These results reflect previously reported effects of NaCl on yeast. Increased amounts of NaCl in the growth environment reduce the number of viable yeast cells as well as the biomass of the culture, while the length of the lag phase is increased [24–27]. The most important metabolite of yeast in bread dough is CO_2, which leavens the dough and increases the volume of the bread loaf; thus, CO_2 can also be used to monitor the yeast activity in dough. Table 2 lists the parameters relating to the yeast performance in wheat dough during fermentation; as can be seen, the total volume of CO_2 produced decreased with increasing amounts of NaCl. The higher the amount of NaCl, the higher the osmotic pressure on the yeast cells, leading to growth inhibition and an inhibitory impact on the yeast metabolism [24,25]. The non-significant difference between bread dough containing 0% and 0.26% NaCl is due to the NaCl threshold concentration required before conditions act to have an inhibitory influence on yeast [27,28]. The higher the NaCl level, the higher the retention coefficient, which leads to a higher retention of CO_2 by the dough, as previously reported [23]. A lower CO_2 production results in a lower pressure on the membranes of the gluten network. In addition, there are several reports that gluten networks are strengthened by NaCl and can thereby retain more CO_2 [23,29,30].

Table 2. Yeast performance in wheat dough with different NaCl concentrations during 3 h fermentation.

NaCl Concentration (% w/w)	Total Volume of Dough (mL)	Volume of Retention (mL)	Volume of CO_2 Lost (mL)	Retention Coefficient
0.00	2241 ± 54 [a]	1435 ± 23 [a]	806 ± 75 [a]	64.1 ± 2.4 [a]
0.26	2108 ± 40 [a]	1459 ± 11 [a]	649 ± 49 [b]	69.2 ± 1.8 [d]
1.04	1581 ± 126 [b]	1337 ± 54 [b]	243 ± 78 [c]	84.8 ± 3.7 [c]
1.73	982 ± 28 [c]	953 ± 22 [c]	30 ± 8 [d]	97.0 ± 0.7 [b]
2.60	573 ± 15 [d]	569 ± 14 [d]	4.0 ± 1.7 [e]	99.4 ± 0.2 [a]
3.46	313 ± 11 [e]	311 ± 11 [e]	2.0 ± 0.0 [f]	99.4 ± 0.1 [a]

Values in one column followed by the same superscript letter are not significantly different ($p < 0.05$).

3.2. Specific Bread Loaf Volume and Bake Loss

The specific volume and bake loss of the baked bread loaves were determined as a standard quality parameter (Table 3). The specific loaf volume decreased significantly for increasing amounts of NaCl at a yeast level of 1.2%. The bread containing no NaCl did not show a significantly larger volume than the bread with 0.3% NaCl, despite its water level being the highest at 62.95% (Table 1).

Table 3. Specific volume and bake loss of bread loaves with different NaCl concentrations and standard yeast level of 1.2% w/w and with different yeast concentrations and standard NaCl concentration of 0.3% w/w.

Bread Type		Specific Bread Volume (mL/g)	Bake Loss *	
			(% w/w)	(% of Water Level)
Varying NaCl (% w/w) at 1.2% w/w yeast	0.0	3.85 ± 0.13 [a]	14.9 ± 0.5 [a,b]	39.1 ± 1.2 [a,b,c]
	0.3	3.81 ± 0.08 [a]	15.3 ± 0.3 [a]	40.6 ± 0.8 [a]
	1.2	3.49 ± 0.09 [b]	14.4 ± 0.4 [b]	38.6 ± 0.9 [b]
	2.0	3.11 ± 0.05 [c]	13.4 ± 0.5 [c]	37.4 ± 0.6 [c]
	3.0	2.52 ± 0.02 [d]	12.3 ± 0.4 [d]	32.7 ± 1.0 [d]
	4.0	1.93 ± 0.03 [e]	10.6 ± 0.3 [e]	28.6 ± 0.8 [e]
Varying yeast (% w/w) at 0.3% w/w NaCl	0.3	3.02 ± 0.02 [i]	13.4 ± 0.2 [g]	36.2 ± 0.5 [g]
	0.6	3.70 ± 0.05 [f]	14.7 ± 0.2 [f]	39.8 ± 0.7 [f]
	0.9	3.62 ± 0.19 [f,g]	14.7 ± 0.2 [f]	39.7 ± 0.6 [f]
	1.2	3.49 ± 0.09 [g,h]	14.4 ± 0.4 [f]	38.6 ± 0.9 [f]
	1.5	3.37 ± 0.10 [h]	13.9 ± 0.5 [f,g]	38.2 ± 0.9 [f]

* The bake loss is shown as a percentage of the dough mass as well as a percentage of the water level. Values in one column followed by the same superscript letter are not significantly different ($p < 0.05$).

Although the dough was mixed to a standard consistency of 500 BU, this observation might be explained by a weaker gluten network in the absence of NaCl [23,29,30]. The data from the rheofermentometer corroborate this hypothesis; the generation of a non-significant higher amount of CO_2 did not lead to a greater bread volume. The higher the NaCl concentration, the smaller was the bread volume and surface area, and less water could evaporate, which again is supported by the rheofermentometer data. The bake loss expressed as a percentage of the water level for each of the individual bread loaves showed that small adjustments of the water level to achieve a dough consistency of 500 BU did not significantly influence the bake loss (Table 3). The breads with yeast levels ranging from 0.3% to 1.5% at a constant NaCl level of 0.3% w/w had the highest specific volume at 0.6%. At lower yeast concentrations, the amount of yeast did not produce sufficient CO_2 to stretch the gluten network to its maximum. At yeast concentrations, higher than 0.6% the specific volume decreased with increasing amounts of yeast. Similarly, increasing amounts of yeast led to excessive fermentation and a resulting expansion of the gluten network, thereby increasing the loss of CO_2 [23,25,27].

3.3. Analyses of Volatile Aroma Compounds in Bread Crumb

Three key aroma compounds in wheat bread crumb are the alcoholic compound 2-phenylethanol [14,31] and the unsaturated aldehydes (E)-2-nonenal and 2,4-(E,E)-decadienal [16,32]. These three compounds were analysed in the aroma extracts of bread crumb by 2D-HRGC-MS (Table 4). The 2-phenylethanol decreased in concentration exponentially with increasing amounts of NaCl, but increased significantly with increasing amounts of yeast, albeit not for the highest yeast concentrations (0.9%, 1.2% and 1.5% w/w).

Table 4. Concentration of 2-phenylethanol, (E)-2-nonenal and 2,4-(E,E)-decadienal in bread-crumb samples as determined by 2D-HRGC-MS.

Bread Type		Concentration (µg/kg)		
		2-Phenylethanol	(E)-2-Nonenal	2,4-(E,E)-Decadienal
Varying NaCl (% w/w) at 1.2% w/w yeast	0.0	4441 ± 686 [a]	12.9 ± 1.7 [c,d]	12.7 ± 1.1 [a]
	0.3	3055 ± 549 [a]	13.5 ± 1.2 [c,d]	11.1 ± 0.7 [a]
	1.2	1582 ± 6 [b]	12.1 ± 0.3 [d]	8.5 ± 0.8 [b]
	2.0	1685 ± 88 [c]	15.9 ± 2.6 [b,c]	9.2 ± 1.3 [a,b]
	3.0	1189 ± 9 [d]	16.4 ± 1.2 [b]	10.6 ± 2.0 [a,b]
	4.0	845 ± 58 [e]	20.8 ± 2.7 [a]	9.3 ± 0.7 [a,b]
Varying yeast (% w/w) at 0.3% w/w NaCl	0.3	1164 ± 64 [i]	19.7 ± 0.0 [e]	20.2 ± 1.8 [c]
	0.6	1773 ± 40 [h]	16.5 ± 1.9 [f]	17.9 ± 3.1 [c]
	0.9	2607 ± 165 [g]	10.2 ± 2.2 [g,h]	11.4 ± 0.7 [d]
	1.2	3056 ± 549 [f,g]	13.5 ± 1.2 [f,g]	11.1 ± 0.7 [d]
	1.5	3288 ± 315 [f]	8.0 ± 2.3 [h]	8.6 ± 0.1 [e]

Values in one column followed by the same superscript letter are not significantly different ($p < 0.05$).

The concentrations of (E)-2-nonenal differed significantly for 3% and 4% NaCl compared to the lowest NaCl concentrations of 0%–0.6%. Increasing the yeast content led to a significant decrease in (E)-2-nonenal for the samples containing 0.3% and 0.6% yeast compared to the samples containing 0.9% and 1.5% yeast. Here, 2,4-(E,E)-decadienal did not show any significant change for the different NaCl concentrations, but decreased with increasing yeast content. These changes can be explained by the reducing activity of the baker's yeast. Decreasing the yeast concentration or inhibiting the yeast with more salt lowers the overall yeast activity during the fermentation process, thereby resulting in a lower production of the unsaturated aldehydes. The reducing activity of yeast has been previously investigated [33]. Furthermore, Saison et al., 2010 demonstrated a change in beer aroma production based on the reducing activity of (E)-2-nonenal by *S. cerevisiae*, suggesting that the reducing activity metabolism of yeast is not affected by increased concentrations of NaCl in the same manner as the Ehrlich pathway or other parts of the yeast metabolism. Correlations ($R^2 \geq 0.75$) were found between CO_2 and 2-phenylethanol or (E)-2-nonenal, indicating that these compounds must arise during yeast metabolism and are influenced by the amount of NaCl in bread dough. A negative correlation for the unsaturated aldehyde (E)-2-nonenal indicates that the reduction capacity of yeast increased with increasing yeast activity, thus an increased amount of (E)-2-nonenal was reduced to the corresponding alcohol.

3.4. PTR-MS Analyses of Bread Dough during Fermentation

The PTR-MS data for the rose-like aroma compound 2-phenylethanol at m/z 105 in the headspace of the dough showed that its concentration increased for increasing amounts of yeast (0.3%–1.5%), albeit with the regressions between 0.6 and 1.2% being similar (Figure 1a). The more NaCl that was added to the bread samples, the less 2-phenylethanol was present due to the inhibiting effect of NaCl on yeast (Figure 1b). The headspace concentrations of 2-phenylethanol correlate with the concentration of this compound in the respective bread crumb samples, as measured by 2D-HRGC-MS.

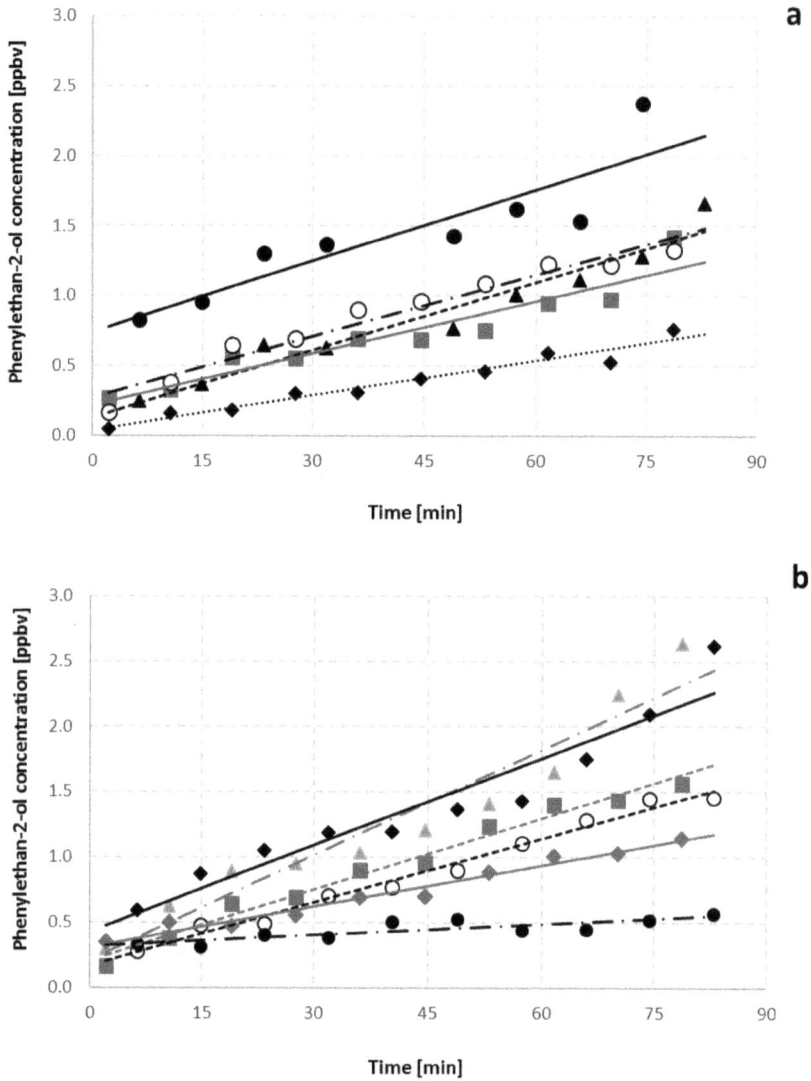

Figure 1. Volume mixing ratios of 2-phenylethanol in the headspace of bread dough containing (a) different yeast concentrations (w/w) of 1.5% (filled circles with solid regression line), 1.2% (open circles with black dot-dashed regression line), 0.9% (filled squares with grey regression line), 0.6% (filled triangles with dashed regression line), 0.3% (filled diamonds with dotted regression line) and (b) different NaCl concentrations (w/w) of 0.0% (filled triangles with grey dot-dashed regression line), 0.3% (filled diamonds with solid regression line), 1.2% (filled squares with grey dashed regression line), 2.0% (open circles with black dashed regression line), 3.0% (filled diamonds with grey regression line) and 4% (filled circles with dot-dashed regression line) over 90 min of incubation under fermentation conditions (30 °C, 85% RH).

3.5. Descriptive Sensory Evaluation

Ten sensory attributes were collected for the bread-crumb samples, whereby the dominating attributes were roasty, yeasty, malty, flour like, buttery, cheesy and fatty. The APAs of the bread-crumb samples revealed significant differences only for the attributes *cheesy* and rose-like (Figure 2).

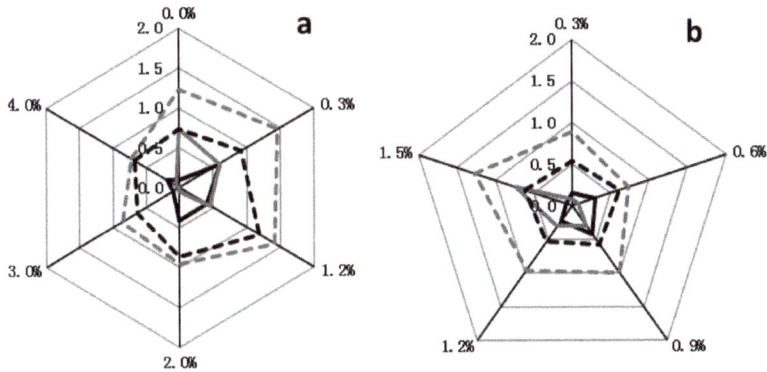

Figure 2. Aroma profiles of bread crumbs for the attributes *cheesy* (grey line), *rose-like* (black line), *fatty* (black dashed line) and *buttery* (grey dashed line) at (**a**) different NaCl concentrations (0.0%, 0.3%, 1.2%, 2.0%, 3.0% and 4% w/w) and (**b**) different yeast levels (0.3%, 0.6%, 0.9%, 1.2% and 1.5% w/w) on a scale from 0 (not detectable) over 1 (weak intensity) to 2 (medium intensity), as determined by sensory panel.

The *cheesy* aroma was significantly different between the samples with the three highest NaCl levels (2%, 3%, and 4% w/w) and the three lowest levels (0.0%, 0.3%, and 1.2% w/w NaCl) (Figure 2a). In the samples with varying yeast levels only the highest (1.5% w/w yeast) and lowest (0.3% w/w yeast) levels differed significantly with respect to the *cheesy* odour impression (Figure 2b). Butanoic acid is well-known as a volatile metabolic compound of yeast with a *cheesy* odour note, and is listed in the Yeast Metabolic Database (YMDB) ID 01392 [34]. The results of these APAs correlate with the yeast activity during fermentation. Increased amounts of yeast as well as decreasing NaCl levels result in higher yeast activities and hence, a higher metabolism rate, including production of the metabolite butanoic acid, which results in a more intense cheesy odour.

The attribute rose-like determined by the sensory panel (Figure 2) correlated directly with the concentration of 2-phenylethanol in the bread crumb samples (Table 4) and hence, with the yeast activity. Increasing yeast activity during dough fermentation led to higher concentrations of 2-phenylethanol. The lower the NaCl level or the higher the yeast level, the more intense was the rose-like odour, as determined by the sensory panel (Figure 2). At 0.0% w/w NaCl the panel did not perceive a rose-like odour, which might reflect a totally excessive and uncontrolled yeast activity in the complete absence of NaCl. Odour impressions described as cheesy and buttery dominated the overall odour characteristic and covered the rose-like impression (Figure 2). The same effect is shown for an increased amount of yeast above 0.9% w/w. The rose-like compounds, namely 2-phenylethanol and 2-phenylacetic acid, have higher odour thresholds compared to the other aroma compounds considered here, therefore the rose-like odour fraction can easily be dominated by other volatile aroma compounds or influence their recognition [22].The fatty impression in the samples with differing NaCl content did not vary significantly, as confirmed by 2D-HRGC-MS analyses of the unsaturated aldehydes (E)-2-nonenal and 2,4-(E,E)-decadienal, both of which have characteristic fatty odour impressions. By contrast, significantly different concentrations of both aldehydes were found in the samples of varying yeast content (Table 4), although the sensory panel could not differentiate between these samples. This might be due to an increasing yeast activity, which predominantly produces other volatile aroma compounds such as butanoic acid and Ehrlich pathway metabolites, which results in an increase in the buttery odour, as observed here (Figure 2).

3.6. Sensory Triangle Test

Sensory triangle test analyses of the two sample pairs with different NaCl levels indicated that there was no significant difference between the pairs ($p > 0.2$). The sample pair with 0.3% and 1.5% w/w yeast was significantly distinguishable ($p < 0.1$). These findings show that NaCl reduction has no significant influence on the overall volatile aroma fraction of the bread crumb recognisable by a human panellist. On the contrary, a 5-fold increase of yeast led to a distinguishable difference based on a 5-fold increase of the yeast metabolism activity in the dough system. The descriptive sensory results show several changes for isolated attributes and compounds in particular based on the determined attributes rose-like and cheesy. However, the aroma profile as a whole did not change to an extent that would be recognisable by consumers.

4. Conclusions

The influence of yeast activity on the generation of volatile aroma compounds in bread crumb was investigated using sensory assessments in combination with two-dimensional high-resolution gas chromatography-mass spectrometry (2D-HRGC-MS) and proton-transfer-reaction mass spectrometry (PTR-MS). A correlation between different yeast metabolites was shown. The metabolic pathways in yeast cells seem to correlate with the reducing activity, independently of the amount of yeast present or the concentration of NaCl. Bread samples with a 5-fold increase in yeast concentration (from 0.3% to 1.5% w/w) were distinguishable by a trained sensory panel. Indeed, when the concentration of NaCl was kept at 0.3%, yeast cells were metabolically over-performing due to the lack of the inhibiting effect normally carried out by the presence of NaCl in the dough system. This related to an increased yeast metabolic activity during the fermentation process, which showed to have a significant impact on the final bread crumb aroma [17]. A reduction in NaCl from the standard concentration of 1.2% w/w to 0.3% w/w increased the yeast activity but the increase in volatile aroma components, as determined by 2D-HRGC-MS and PTR-MS, could not be detected by the sensory panel. These observations suggest that NaCl reduction does not influence the volatile aroma of bread significantly and consumers in general are not able to recognise any reduced amounts of salt in the odour of bread crumb. While salt reduction in bread impacts on the quality characteristics of taste, shelf-life and texture [22,35], the aroma quality remains unchanged.

Acknowledgments: The authors would like to thank Franziska Wiegand and Tom Hannon for their support. The authors wish to acknowledge that this project was funded under the Irish National Development Plan, through the Food Institutional Research Measure, administered by the Department of Agriculture, Fisheries and Food, Ireland.

Author Contributions: All authors participated in the experimental design of the study. Emanuele Zannini and Elke Arendt coordinated the execution of the study. Jonathan Beauchamp and Michael Czerny contributed their knowledge, expertise and support in the 2D-HRGC-MS, PTR-MS and sensory analysis. Markus Belz performed the experiments, data processing and drafted the manuscript. Claudia Axel, Emanuele Zannini and Jonathan Beauchamp corrected and revised the manuscript to its final version.

Conflicts of Interest: The authors declare no conflict of interest.

References

1. Tuorila-Ollikainen, H.; Lahteenmaki, L.; Salovaara, H. Attitudes, norms, intentions and hedonic responses in the selection of low salt bread in a longitudinal choice experiment. *Appetite* **1986**, *7*, 127–139. [CrossRef]
2. Wyatt, C. Acceptability of Reduced Sodium in Breads, Cottage Cheese, and Pickles. *J. Food Sci.* **1983**, *48*, 1300–1302.
3. Bassett, M.N.; Pérez-Palacios, T.; Cipriano, I.; Cardoso, P.; Ferreira, I.M.P.L.V.O.; Samman, N.; Pinho, O. Development of Bread with NaCl Reduction and Calcium Fortification: Study of Its Quality Characteristics. *J. Food Qual.* **2014**, *37*, 107–116. [CrossRef]

4. Charlton, K.E.; MacGregor, E.; Vorster, N.H.; Levitt, N.S.; Steyn, K. Partial replacement of NaCl can be achieved with potassium, magnesium and calcium salts in brown bread. *Int. J. Food Sci. Nutr.* **2007**, *58*, 508–521. [CrossRef] [PubMed]
5. Rizzello, C.G.; Nionelli, L.; Coda, R.; Di Cagno, R.; Gobbetti, M. Use of sourdough fermented wheat germ for enhancing the nutritional, texture and sensory characteristics of the white bread. *Eur. Food Res. Technol.* **2010**, *230*, 645–654. [CrossRef]
6. Ghawi, S.K.; Rowland, I.; Methven, L. Enhancing consumer liking of low salt tomato soup over repeated exposure by herb and spice seasonings. *Appetite* **2014**, *81*, 20–29. [CrossRef] [PubMed]
7. Jimenez-Maroto, L.A.; Sato, T.; Rankin, S.A. Saltiness potentiation in white bread by substituting sodium chloride with a fermented soy ingredient. *J. Cereal Sci.* **2013**, *58*, 313–317. [CrossRef]
8. Kuo, W.-Y.; Lee, Y. Effect of Food Matrix on Saltiness Perception—Implications for Sodium Reduction. *Compr. Rev. Food Sci. Food Saf.* **2014**, *13*, 906–923. [CrossRef]
9. Pflaum, T.; Konitzer, K.; Hofmann, T.; Koehler, P. Influence of Texture on the Perception of Saltiness in Wheat Bread. *J. Agric. Food Chem.* **2013**, *61*, 10649–10658. [CrossRef] [PubMed]
10. Noort, M.W.J.; Bult, J.H.F.; Stieger, M. Saltiness enhancement by taste contrast in bread prepared with encapsulated salt. *J. Cereal Sci.* **2012**, *55*, 218–225. [CrossRef]
11. Schieberle, P.; Grosch, W. Potent odorants of the wheat bread crumb differences to the crust and effect of a longer dough fermentation. *Z. Lebensm.-Unters. Forsch.* **1991**, *192*, 130–135. [CrossRef]
12. Suomalainen, H.; Lehtonen, M. The production of aroma compounds by yeast. *J. Inst. Brew.* **1979**, *85*, 149–156. [CrossRef]
13. Whiting, G.C. Organic acid metabolism of yeasts during fermentation of alcoholic beverages—A review. *J. Inst. Brew.* **1976**, *82*, 84–92. [CrossRef]
14. Czerny, M.; Schieberle, P. Labelling studies on pathways of amino acid related odorant generation by *Saccharomyces cerevisiae* in wheat bread dough. In *Flavour Science: Recent Advances and Trends*; Bredie, W.L.P., Petersen, M.A., Eds.; Elsevier: Amsterdam, The Netherland, 2006; pp. 89–92.
15. Saison, D.; De Schutter, D.P.; Vanbeneden, N.; Daenen, L.; Delvaux, F.; Delvaux, F.R. Decrease of aged beer aroma by the reducing activity of brewing yeast. *J. Agric. Food Chem.* **2010**, *58*, 3107–3115. [CrossRef] [PubMed]
16. Vermeulen, N.; Czerny, M.; Gänzle, M.G.; Schieberle, P.; Vogel, R.F. Reduction of (E)-2-nonenal and (E,E)-2,4-decadienal during sourdough fermentation. *J. Cereal Sci.* **2007**, *45*, 78–87. [CrossRef]
17. Birch, A.N.; Petersen, M.A.; Arneborg, N.; Hansen, Å.S. Influence of commercial baker's yeasts on bread aroma profiles. *Food Res. Int.* **2013**, *52*, 160–166. [CrossRef]
18. Birch, A.N.; Petersen, M.A.; Hansen, Å.S. The aroma profile of wheat bread crumb influenced by yeast concentration and fermentation temperature. *LWT Food Sci. Technol.* **2013**, *50*, 480–488. [CrossRef]
19. Martins, Z.E.; Erben, M.; Gallardo, A.E.; Barbosa, I.; Pinho, O.; Ferreira, I.M.P.L.V.O. Effect of spent yeast fortification on physical parameters, volatiles and sensorial characteristics of home-made bread. *Int. J. Food Sci. Technol.* **2015**, *50*, 1855–1863. [CrossRef]
20. Engel, W.; Bahr, W.; Schieberle, P. Solvent assisted flavour evaporation—A new and versatile technique for the careful and direct isolation of aroma compounds from complex food matrices. *Eur. Food Res. Technol.* **1999**, *209*, 237–241. [CrossRef]
21. Lindinger, W.; Hansel, A.; Jordan, A. Proton-transfer-reaction mass spectrometry (PTR-MS): On-line monitoring of volatile organic compounds at pptv levels. *Chem. Soc. Rev.* **1998**, *27*, 347–375. [CrossRef]
22. Czerny, M.; Christlbauer, M.; Christlbauer, M.; Fischer, A.; Granvogl, M.; Hammer, M.; Hartl, C.; Hernandez, N.M.; Schieberle, P. Re-investigation on odour thresholds of key food aroma compounds and development of an aroma language based on odour qualities of defined aqueous odorant solutions. *Eur. Food Res. Technol.* **2008**, *228*, 265–273. [CrossRef]
23. Lynch, E.J.; Dal Bello, F.; Sheehan, E.M.; Cashman, K.D.; Arendt, E.K. Fundamental studies on the reduction of salt on dough and bread characteristics. *Food Res. Int.* **2009**, *42*, 885–891. [CrossRef]
24. Almagro, A.; Prista, C.; Castro, S.; Quintas, C.; Madeira-Lopes, A.; Ramos, J.; Loureiro-Dias, M.C. Effects of salts on *Debaryomyces hansenii* and *Saccharomyces cerevisiae* under stress conditions. *Int. J. Food Microbiol.* **2000**, *56*, 191–197. [CrossRef]
25. Watson, T.G. Effects of sodium chloride on steady-state growth and metabolism of *Saccharomyces cerevisiae*. *J. Gen. Microbiol.* **1970**, *64*, 91–99. [CrossRef] [PubMed]

26. Wei, C.-J.; Tanner, R.D.; Malaney, G.W. Effect of Sodium Chloride on Bakers' Yeast Growing in Gelatin. *Appl. Environ. Microbiol.* **1982**, *43*, 757–763. [PubMed]

27. Oda, Y.; Tonomura, K. Sodium chloride enhances the potential leavening ability of yeast in dough. *Food Microbiol.* **1993**, *10*, 249–254. [CrossRef]

28. Kawai, H.; Bagum, N.; Teramoto, I.; Isobe, U.; Yokoigawa, K. Effect of sodium chloride on the growth and fermentation activity of Saccharomyces yeasts. *Res. J. Living Sci.* **1999**, *45*, 55–61.

29. Dal Bello, F.; Clarke, C.I.; Ryan, L.A.M.; Ulmer, H.; Schober, T.J.; Ström, K.; Sjögren, J.; van Sinderen, D.; Schnürer, J.; Arendt, E.K.; et al. Improvement of the quality and shelf life of wheat bread by fermentation with the antifungal strain *Lactobacillus plantarum* FST 1.7. *J. Cereal Sci.* **2007**, *45*, 309–318. [CrossRef]

30. Beck, M.; Jekle, M.; Becker, T. Sodium chloride—Sensory, preserving and technological impact on yeast-leavened products. *Int. J. Food Sci. Technol.* **2012**, *47*, 1798–1807. [CrossRef]

31. Ehrlich, F. Über die Bedingungen der Fuselölbildung und über ihren Zusammenhang mit dem Eiweißaufbau der Hefe. *Ber. Deutsch. Chem. Ges.* **1907**, *40*, 1027–1047. [CrossRef]

32. Frasse, P.; Lambert, S.; Levesque, C.; Melcion, D.; Richard-Molard, D.; Chiron, H. The influence of fermentation on volatile compounds in French bread crumb. *Food Sci. Technol.—Lebensm.-Wiss. Technol.* **1992**, *25*, 66–70.

33. Chwastowski, J.; Koloczek, H. The kinetic reduction of Cr (VI) by yeast *Saccharomyces cerevisiae*, *Phaffia rhodozyma* and their protoplasts. *Acta Biochim. Pol.* **2013**, *60*, 829–834. [PubMed]

34. Jewison, T.; Knox, C.; Neveu, V.; Djoumbou, Y.; Guo, A.C.; Lee, J.; Liu, P.; Mandal, R.; Krishnamurthy, R.; Sinelnikov, I.; et al. YMDB: The Yeast Metabolome Database. *Nucl. Acids Res.* **2012**, *40*, D815–D820. [CrossRef] [PubMed]

35. Belz, M.C.E.; Mairinger, R.; Zannini, E.; Ryan, L.A.M.; Cashman, K.D.; Arendt, E.K. The effect of sourdough and calcium propionate on the microbial shelf-life of salt reduced bread. *Appl. Microbiol. Biotechnol.* **2012**, *96*, 493–501. [CrossRef] [PubMed]

foods

MDPI

Article

The Content of Tocols in South African Wheat; Impact on Nutritional Benefits

Maryke Labuschagne [1], Nomcebo Mkhatywa [1], Eva Johansson [2,*], Barend Wentzel [3] and Angeline van Biljon [1]

[1] Department of Plant Sciences, University of the Free State, Bloemfontein 9300, South Africa; LabuscM@ufs.ac.za (M.L.); 2006041384@ufs4life.ac.za (N.M.); avbiljon@ufs.ac.za (A.v.B.)

[2] Department of Plant Breeding, The Swedish University of Agricultural Sciences, Box 101, SE-230 53 Alnarp, Sweden

[3] Small Grains Institute, Bethlehem 9700, South Africa; wentzelb@arc.agric.za

* Correspondence: Eva.johansson@slu.se; Tel: +46-40-415-562; Fax: +46-40-415-519

Received: 31 August 2017; Accepted: 27 October 2017; Published: 2 November 2017

Abstract: Wheat is a major component within human consumption, and due to the large intake of wheat, it has an impact on human nutritional health. This study aimed at an increased understanding of how the content and composition of tocols may be governed for increased nutritional benefit of wheat consumption. Therefore, ten South African wheat cultivars from three locations were fractionated into white and whole flour, the content and concentration of tocols were evaluated by high performance liquid chromatography (HPLC), and vitamin E activity was determined. The content and composition of tocols and vitamin E activity differed with fractionation, genotype, environment, and their interaction. The highest tocol content (59.8 mg kg^{-1}) was obtained in whole flour for the cultivar Elands grown in Ladybrand, while whole Caledon flour from Clarence resulted in the highest vitamin E activity (16.3 mg kg^{-1}). The lowest vitamin E activity (1.9 mg kg^{-1}) was found in the cultivar C1PAN3118 from Ladybrand. High values of tocotrienols were obtained in whole flour of the cultivars Caledon (30.5 mg kg^{-1} in Clarens), Elands (35.5 mg kg^{-1} in Ladybrand), and Limpopo (33.7 mg kg^{-1} in Bultfontein). The highest tocotrienol to tocopherol ratio was found in white flour (2.83) due to higher reduction of tocotrienols than of tocopherols at fractionation. The quantity and composition of tocols can be governed in wheat flour, primarily by the selection of fractionation method at flour production, but also complemented by selection of genetic material and the growing environment.

Keywords: *Triticum aestivum*; tocopherol; tocotrienol; vitamin E; genotype; environment

1. Introduction

Wheat is, together with rice, the major food crop in the world, supporting 20% of the daily energy for the human population [1]. In certain parts of the world, wheat is even the main staple food, contributing up to 70% of the daily energy and protein in the human diet [2]. Due to its high consumption, wheat provides a significant amount of energy, proteins, and selected micronutrients and vitamins to the consumer [3–5]. Thus, despite the fact that food from other origins might to a large extent have a higher relative content of certain compounds, wheat serves as a source of important nutritional components such as iron and zink, vitamin E, phenolics, and carotenoids [6]. The content of vitamin E and its activity is determined by the content of certain tocols [7]. The tocols are known as bioactive compounds with antioxidant traits, for e.g., they prevent oxidation of double bonds by reacting with peroxyl radicals and protecting lipids and membrane proteins against oxidative stress [8,9]. Tocols cannot be produced by humans and are, therefore, important components of any diet, and they are obtained from a large number of plant based foods [6]. Tocols consists of eight lipid-soluble compounds: α-, β-, γ-, δ-tocotrienol, and α-, β-, γ-, δ-tocopherol. Previous literature has accounted different vitamin E activity to the various tocopherols due to their chemical structures and physiological factors. However, recent opinion considers only α-tocopherol as the source of vitamin E activity [7]. For whole wheat flour, a mean vitamin E content of 0.71 mg/100 g has been reported by the U.S. Department of Agriculture (USDA) [10]. However, higher levels of vitamin E have been reported in whole wheat flour from certain genetic material, resulting in the presence of 20% of the daily requirement of vitamin E in 200 g of whole wheat flour [11]. Poor nutritional status together with a high prevalence of stressors, for e.g., malaria and HIV as may be the status of many in developing countries, are known to contribute towards vitamin E deficiency [12]. Thus, selection of suitable wheat material with high vitamin E content might contribute to solving the problems of vitamin E deficiency. Tocotrienols do not show vitamin E activity but are known to have higher antioxidant activity than the tocopherols and also have additional important health promoting effects [11,13]. Tocopherols are widely distributed in higher plants whereas tocotrienols occur mainly in some non-photosynthetic tissues such as seeds and endosperm of monocot grains [14]. In the wheat grain, α- and β-tocopherols are mainly found in the wheat germ, while tocotrienols are concentrated in the pericarp, testa, aleurone, and in the endosperm.

Fractionation of the crushed grain during milling is known to have a critical implication for the distribution of many nutrients [15]. The consumption of whole grain is a healthy alternative to white flour [16]. Previous studies have also shown genotype, environment, and cultivation practices to have an impact on tocol content and composition [6,11,14]. Increased understanding of the impact and interactions of genotype, environment, and processing on content of tocols and vitamin E activity in wheat flour will positively impact a healthy intake of these compounds from wheat.

Thus, the aim of this study was to evaluate the effects of genotype, environment, and fractionation through milling on the quantity and composition of tocol components in South African wheat flour. Interactions as well as importance of the various evaluated factors will be investigated and conclusions drawn as related to opportunities to govern these health components in wheat flour.

2. Materials and Methods

2.1. Plant Material

Ten South African bread wheat cultivars, Betta-DN, Caledon, Elands, Gariep, Komati, Limpopo, Matlabas, PAN3118, PAN3349, and PAN3377, grown in four replicates in each location, were used in the present study. The trials were conducted at three different locations in one season: Bultfontein (28°16′53.14″ S 26°27′02.77″ E, north western Free State with low rainfall, high temperatures, high evaporation requirements and deep, yellow sandy loam soils with a water table present), Ladybrand (29°14′30.75″ S 27°20′18.55″ E, central Free State, moderate rainfall, moderate temperatures, a lower evaporation requirement and relatively shallow duplex soils), and Clarens (28°24′26.63″ S 27°20′18.55″ E, eastern Free State, higher rainfall, lower temperatures, lower evaporation requirement with predominantly yellow soils of average effective depth). The trials were planted under dryland conditions in a randomized complete block design. Trial plots consisted of five rows of 5 m length and inter-row spacing of 5 cm. Fertilization was done after soil analysis and according to normal production practices for each location.

2.2. Extraction of Tocols

Milled grain samples were freeze dried for three days before tocol extraction. The extraction [17] was performed with modifications as suggested by Labuschagne et al. [18].

2.3. Analytical High Performance Liquid Chromatography (HPLC)

A normal phase-HPLC method [19] with modification [18] was used to separate the tocol compounds (Figure 1). A Phenomenex Luna Silica column (250 mm × 4.6 mm inner diameter (i.d.), 5 µm particle size) was used. The mobile phase was *n*-hexane/ethyl acetate/acetic acid (97.3:1.8:0.9 *v/v/v*) at a flow rate of 1.6 mL min^{-1}. All peaks were detected by fluorescence and the wavelength of detection was set to 290 nm and emission wavelength of 330 nm. HPLC injection volume was 10 µL per injection. A standard solution was used to carry out the linearity test over the different concentration ranges (ng µL^{-1}) close to the amount of tocols found in the samples: α-tocopherol 0.47–9.57 ng µL^{-1}; β-tocopherol 0.23–4.7 ng µL^{-1}; γ-tocopherol 0.65–13.1 ng µL^{-1}; δ-tocopherol 0.62–12.4 ng µL^{-1}; β-tocotrienol 0.54–10.82 ng µL^{-1}. Total tocols were the sum of α-tocopherol, β-tocopherol, α-tocotrienol, β-tocotrienol, and δ-tocotrienol.

2.4. Data Analysis

Statistical evaluation applying ANOVA followed by mean comparison with Duncan post-hoc test at *p* < 0.05, Spearman rank correlation analyses, and Principal component analyses (PCA) was carried out using the statistical package SAS (2004; SAS Institute Inc., Cary, NC, USA). In order to explain the proportion of the contribution of variation by the environments, genotypes, and flour fractionation on the tocol composition, regression analysis was applied [20,21]. Vitamin E activity was calculated based on α-tocopherol content according to the Scientific Opinion of an EFSA (European Food Safety Authority) Panel [7]. The Recommended Daily Intake (RDI) of vitamin E set by European Parliament and the Council in the Regulation No. 1169/2011 of 25 October 2011 is 12 mg/day [22].

3. Results

3.1. Importance of Genotype, Environment, and Fractionation on Tocols Content and Composition

The percentage recovery of tocols was more than 95%, and the different tocols were successfully separated by HPLC (Figure 1). The major tocols found were α- and β-tocopherol and α- and β-tocotrienol (Figure 1). Delta-tocopherols and especially γ-tocopherols were only found in small amounts and often only in traces. Flour type (white flour versus whole flour) was shown to explain by far the highest part of the variation in tocols content and composition except for δ-tocotrienol, where location was a significant parameter for the variation (Table 1). However, combination of flour type, cultivar, and location resulted in a higher degree of explanation as compared to each of the factors alone (Table 1), and analysis of variance (ANOVA) also showed significant interactions among the factors for content and composition of the tocols (*p* < 0.01). Similarly, the PCA analysis showed samples clearly clustering in two separate groups as related to flour type, although a number of whole flour samples from Ladybrand were also diverging into a separate group of whole flour samples based on lower values on the second principal component value (Figure 2).

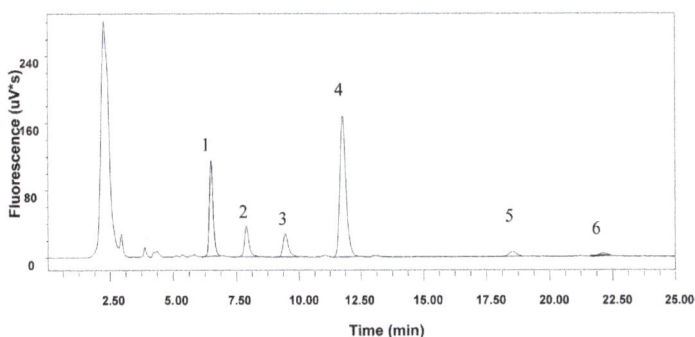

Figure 1. Example of separation of tocols by HPLC (High Performance Liquid Chromatography) from one wheat sample. Peak 1 = α-tocopherol, Peak 2 = α-tocotrienol, Peak 3 = β-tocopherol, Peak 4 = β-tocotrienol, Peak 5 = δ-tocopherol, and Peak 6 = δ-tocotrienol.

Table 1. Percentage of explanation (obtained through R-square from simple linear regression) of flour types (Flour; F), varieties (V), and growing locations (L) as well as their combinations on various tocols.

	α-TP	β-TP	α-TT	β-TT	δ-TT	TP	TT	Tot
Flour	89.0	89.6	87.8	83.5	11.2	89.9	85.8	90.3
Variety	1.06	1.45	3.53	1.46	4.29	1.18	2.06	1.64
Location	0.90	0.01	0.46	1.38	15.1	0.49	1.16	0.70
F, V, L	90.9	91.1	91.8	86.4	30.7	91.6	89.0	92.6

TP = tocopherols, TT = tocotrienols, Tot = total tocols.

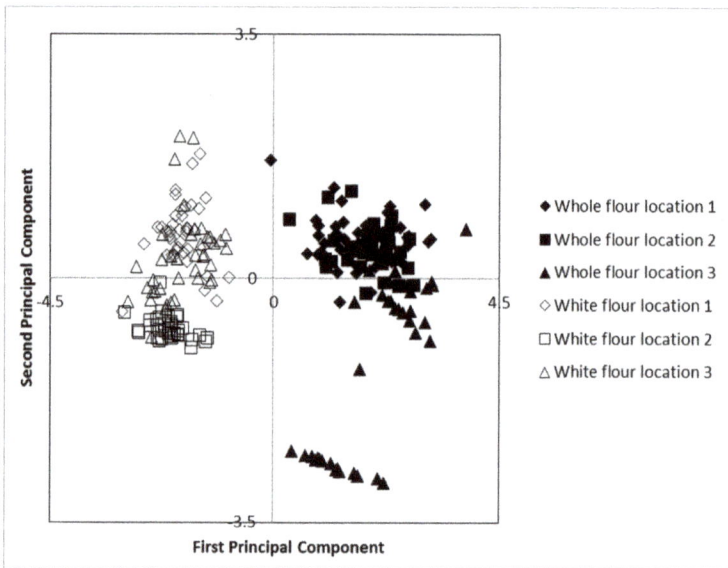

Figure 2. Loading plot from principal component analysis of tocopherols in two flour types of ten varieties grown at three locations in South Africa. Location 1 = Bultfontein, Location 2 = Clarens, Location 3 = Ladybrand. First principal component explained 79.3% of the variation while the second principal component explained 17.4% of the variation.

3.2. Effect of Flour Type on Content and Composition of Tocols

Whole flour had significantly ($p < 0.005$) higher tocol concentrations than white flour, the latter having on average 40% of the concentration of whole flour (Table 2). Concentration in white flour of α-tocopherol was 24% of that in whole flour and concentration in white flour of β-tocopherol was 31% of that in whole flour. Furthermore, white flour contained 25% of the α-tocotrienol concentration found in whole flour and 53% of the β-tocotrienols. The fact that the β-tocotrienols were retained to a higher degree than the other tocols from the whole to the white flour could also be seen as a higher tocotrienol to tocopherol quota (TT/TP) in the white flour as related to whole flour (Table 2). Tocol concentration in white flour as a percentage of that in whole flour varied in genotypes and locations, from 31% (PAN 3349 in Ladybrand) to 51% (Caledon at Bultfontein; Table 3).

Table 2. Mean values (mg kg^{-1}) of various tocols depending on flour type (white versus whole meal flour).

Flour	α-TP	β-TP	α-TT	β-TT	δ-TT	TT/TP	Tot
White	3.34 [b]	2.00 [b]	1.24 [b]	12.7 [b]	0.45 [b]	2.83 [a]	19.7 [b]
Whole	13.3 [a]	6.43 [a]	4.93 [a]	24.0 [a]	0.60 [a]	1.49 [b]	49.4 [a]

Average values followed by the same letters are not significantly different at $p < 0.05$ applying Duncan post-hoc test. TP = tocopherols, TT = tocotrienols, Tot = total tocols.

Table 3. Total tocol content (mg kg^{-1}) mean values of ten cultivars in three locations in white and whole flour.

Flour Type	Cultivar	Bultfontein	Clarens	Ladybrand	Average
	Betta-DN	21.2	17.5	19.0	19.3 [cd]
	Caledon	26.4	24.1	25.9	25.5 [a]
	Elands	16.4	20.6	23.8	20.3 [bc]
	Gariep	20.1	20.0	24.7	21.6 [b]
	Komati	20.6	21.4	23.6	21.9 [b]
White	Limpopo	17.9	20.7	17.4	18.6 [cde]
	Matlabas	18.8	18.7	18.3	18.6 [cde]
	C1PAN3118	17.1	17.0	15.3	16.5 [f]
	C2PAN3349	18.0	16.6	15.9	16.8 [f]
	C3PAN3377	18.2	17.8	17.8	18.1 [def]
	Average	19.5 [a]	19.4 [a]	20.2 [a]	
	Betta-DN	48.0	51.2	52.8	50.6 [bc]
	Caledon	52.0	55.0	57.6	55.0 [a]
	Elands	49.9	51.4	59.8	53.7 [a]
	Gariep	40.9	48.6	49.6	46.3 [d]
	Komati	43.8	51.8	52.3	49.3 [bc]
Whole	Limpopo	55.7	49.5	55.7	53.7 [a]
	Matlabas	39.4	52.8	52.0	48.2 [cd]
	C1PAN3118	43.5	42.9	45.4	43.9 [e]
	C2PAN3349	40.4	41.7	43.6	41.9 [e]
	C3PAN3377	50.2	47.7	54.5	51.0 [b]
	Average	46.5 [c]	49.2 [b]	52.3 [a]	

Average values followed by the same letters are not significantly different at $p < 0.05$ applying Duncan post-hoc test.

3.3. Effect of Cultivar and Growing Location on Content and Composition of Tocols

Tocol concentration in genotypes ranged between 16.49–25.49 mg kg^{-1} for white flour and 41.92–54.87 mg kg^{-1} for whole flour (Table 3). Thus, significant differences in tocol content and composition were found among certain of the evaluated cultivars (Table 3), despite the fact that only a limited amount of variation was explained by the differentiation in cultivars (Table 1 and Figure 2). Among the evaluated cultivars, Caledon was found to have a high concentration of tocols in whole flour and the highest concentration among the cultivars in white flour.

Cultivation of the cultivars in Ladybrand resulted in a higher tocol concentration in whole flour and a higher concentration in the white flour as compared to that of the other localities, with an average of 20.1 mg kg^{-1} for white flour and 52.3 mg kg^{-1} for whole flour (Table 3). In general, total tocol content in the samples differed significantly when grown at the different localities for whole flour but no such significant differences were found for white flour. Part of the explanation for this variation in the whole flour (Figure 2) might be the presence of δ-tocotrienol in samples grown in Ladybrand which were only found in trace amounts in samples from the other growing locations. However, a significant variation was found for the various tocol compound concentrations in both white and whole flour from the different locations. Samples from Clarens and Ladybrand had a significantly higher concentration of α-tocopherol in white and whole flour and β-tocopherol in white flour as compared to samples from Bultfontein, while samples from Clarence were higher than those from the other localities for β-tocopherol in whole flour. Significantly higher concentrations of α-tocotrienol and β-tocotrienol were found in whole flour samples from Ladybrand as compared to samples from the other locations, while for white flour, samples from Ladybrand and Bultfontein showed higher concentrations of α-tocotrienol than samples from Clarence (Table 4). In general, samples in Ladybrand showed a relatively high content of both tocopherols and tocotrienols in both white and whole flour, while samples from Clarence showed a relatively high tocotrienol concentration in both types of flours. Samples from Bultfontein showed a relatively high content of tocotrienols only in white flour. Due to differences in variations of concentrations of various tocol compounds by location, the tocotrienol to tocopherol quota (TT/TP) of the samples differed among locations; with significantly higher values in both white and whole flour samples from Ladybrand and Bultfontein as compared to those from Clarence (Table 4).

Table 4. Average content of each tocol compound (mg kg^{-1}) in three locations from white and whole wheat.

Flour Type	Characteristic	Bultfontein	Clarens	Ladybrand
White	α-Tocopherol	3.02 [b]	3.59 [a]	3.36 [a]
	β-Tocopherol	1.88 [b]	2.08 [a]	2.02 [a]
	α-Tocotrienol	1.32 [a]	1.12 [b]	1.28 [a]
	β-Tocotrienol	12.7 [a]	12.4 [a]	13.0 [a]
	TT/TP	3.07 [a]	2.51 [b]	2.92 [a]
Whole	α-Tocopherol	11.8 [b]	14.1 [a]	14.0 [a]
	β-Tocopherol	6.30 [b]	6.62 [a]	6.33 [b]
	α-Tocotrienol	4.80 [b]	4.45 [c]	5.52 [a]
	β-Tocotrienol	22.7 [c]	23.4 [b]	26.0 [a]
	TT/TP	1.53 [a]	1.36 [b]	1.57 [a]

TT = Tocotrienol, TP = Tocopherols; Average values of the different characteristics followed by the same letters are not significantly different at $p < 0.05$ applying Duncan post-hoc test.

4. Discussion

The present study clearly shows that the quantity and composition of tocols can be governed in South African wheat flour; fractionation through milling is the major determinant, but the effect was found to interact with the selection of genetic material and the growing environment. As can be seen from calculations of daily requirements of vitamin E, only 5% on average was obtained from white South African wheat flour while 22% was obtained from the corresponding whole flour (Table 5). However, by selecting whole flour of a high vitamin E cultivar (Caledon) grown on the locality (Clarence) contributing most to a high level of vitamin E activity, 27% of the daily requirement could be obtained by consumption of 200 g wheat flour per day (Table 5). Tocols are known to be destroyed when heated (25–94% reduction in vitamin E activity) [23,24]. A recent investigation showed a 40% reduction in tocopherols in bread as compared to their corresponding flour, although toasting of the bread resulted in an increase in tocopherol content so that the content reached 89% of the original content of the flour [25]. Today, most wheat is consumed after a heat treatment, thereby reducing the tocol content; consumption of wheat as whole and/or sprouted grain products is the best solution for wheat to be a tocol source [11]. However, the findings that toasting increases the amount of tocopherols as related to what is found in the bread calls for additional evaluations of how to best process wheat products in order to use these products as tocol sources [25]. To secure a high intake of tocols from the food, flour based products should be combined with other food items very high in tocols content [6,26,27].

Table 5. Vitamin E activity of wheat flour of different fractions, genotypes, and localities, calculated as tocopherol equivalents [7] and percentage of recommended intake [11,19] from the average flour consumption in the world of 200 g/person/day.

Flour Origin	Vitamin E Activity (mg kg^{-1})	Recommended Daily Intake (mg)	Percentage of Recommended Vitamin E from 200 g of Wheat Flour (%)
White	3.3	12	5.5
Whole	13.3	12	22.2
Whole flour at various locations			
Bultfontein	11.9	12	19.8
Clarence	14.1	12	23.5
Whole flour from Clarence in various cultivars			
Caledon	16.1	12	26.8
C1PAN3118	12.3	12	20.5

Vitamin E activity is primarily based on the presence of tocopherols in the food [24,28–30]. Besides variation in vitamin E activity, a large variation was also noted in the quantity and composition of tocotrienols in the present wheat material, and this related to variation in fractionation through milling, genotype, and location. Several investigations and results have indicated that tocotrienols are potentially as important for human health as are tocopherols. The tocotrienols have been found to have higher antioxidant capacity [31] and different health promoting properties than those of tocopherols [32–35]. Examples of biological tocotrienols-mediated activities not shared with the tocopherols are neuroprotection, radio-protection, anti-cancer, anti-inflammatory, and liquid lowering properties [36–39]. Similar to what was found for vitamin E activity, fractionation was by far the most important factor for the quantity of the tocotrienols found in the flour. However, tocotrienols content decreased less than tocols content with milling to white flour, resulting in a higher tocotrienols to tocols content in the white flour as compared to that of whole flour. Also, cultivation location resulting in the highest tocotrienols content varied for tocotrienols as compared to vitamin E activity. For tocotrienols, Ladybrand resulted in the highest amount among the locations with Caledon, Limpopo, and Elands showing the highest values among the cultivars. Scientifically based recommendations as to daily intake are only available for vitamin E [11] and not for the separate isoforms, although retailers are announcing that 34–43 mg per day of tocotrienols are beneficial [40]. If such a level should be recommended, the present white and whole flour samples would contribute to around 6% and 12%, respectively, of the daily requirements at a consumption of 200 g of flour per day. However, recent literature reports highly divergent bioavailability of tocotrienols based on the source of tocotrienols and on the target population [36]. Various sources of tocotrienols are rich to different degrees in different tocotrienol compounds, for e.g., tocotrienols from palm oil (the most common source of tocotrienol supplements) is particularly rich in δ-tocotrienol, while β-tocotrienol was the predominant compound in the present material, as also shown in previous studies on wheat [41,42]. The fact that the composition of tocotrienols varies with the source of tocotrienols may be one explanation for the difficulties of making recommendations for the daily intake requirements. Also, it is unclear what effect the relationship of the various tocotrienols and the ratio between tocotrienols and tocopherols is playing. Similar to previous studies [43], we were able in this study to show that cultivar and cultivation location affect composition of tocotrienols, for e.g., cultivation in Ladybrand did not only result in high levels of tocotrienols but also in higher levels of δ-tocotrienol in the whole flour than cultivation in the other locations. Ladybrand has a moderate rainfall and temperatures, resulting in lower evaporation and a relatively shallow duplex soil, growing parameters that might be the background for the content and composition of tocols. However, relationships among growing parameters and tocols content and composition have to be further evaluated before conclusions are made.

Scientific literature recommends not setting daily requirements of tocotrienols until more research based evidence is available [44]. However, from our study, we can conclude that fractionation, cultivation place, and cultivar need to be taken into consideration together with processing of the food in order to determine health effects of a certain food source of tocotrienols.

An increased understanding of how various tocol compounds can be governed through various production factors, including fractionation, genotype, and locality, opens up opportunities of producing specifically useful raw material for further food processing into nutritional beneficial food products with high bioavailability of wanted compounds. Fractionation was the factor effecting quantity of all tocols to the highest extent. In general, grain antioxidants are known to be concentrated in the bran and germ fractions, and thus these fractions are the major contributors to the total antioxidant activities of wheat [45,46]. A number of studies have also shown significantly more antioxidant activity in products manufactured with whole grains than products from refined wheat [47,48]. Previous studies have indicated that tocopherols are more concentrated in the germ fraction, while tocotrienols are found more in bran and are also observed in the endosperm [16,49]. This agrees with the findings in this study, which showed a higher content of all tocols in the whole flour compared to the white flour, but with the reduction to different extents depending on the compound. Fractionation procedures might, therefore, be a useful tool to effectively modulate tocol content and composition in wheat flour, although it can be complimented with genetic impact through the choice of cultivars and environmental impact through the choice of cultivation location.

5. Conclusions

The quantity and composition of tocols can be governed in wheat flour, primarily by the selection of the fractionation method at flour production, but also complemented by selection of genetic material and the growing environment. Total high content of tocopherols in the flour can best be obtained by the selection of whole flour from a high tocol producing cultivar cultivated in good conditions (not hot and dry). By doing so, 27% of the human daily requirement of vitamin E can be received by an average consumption of 200 g of wheat flour per day unless the tocols in the wheat flour are destroyed by harsh processing conditions.

Tocotrienols are possibly even more important than vitamin E in their contribution from wheat towards a healthy diet. Content and composition of these compounds can also be governed in wheat, as large variations are present due to fractionation, cultivar, and cultivation location. However, to do so, an increased knowledge on the requirements as to total amount, composition, and ratios of the different compounds is needed.

Acknowledgments: This study was supported by the Swedish University of Agricultural Sciences (SLU) and the UD-40 project (Ministry of Foreign Affairs in Sweden administered through the SLU). The National Research Foundation (project UID72056) is also acknowledged for financial support. Thanks to Maria Luisa Prieto-Linde for technical assistance with sample analyses at SLU.

Author Contributions: All authors planned the manuscript jointly and participated in discussions of the manuscripts as well as commented on the various drafted text versions of the manuscript. The first author (M.L.) came up with the idea of the manuscript and also completed the main part of the writing and compilation of various parts of the manuscript together with the third author (E.J.). The rest of the authors took the main responsibility for different parts of the experiments; N.M. carried out most of the lab work and a first compilation of the results of the data, B.W. took the main responsibility for the plant material in terms of its selection and cultivation, A.B. was responsible for analyses carried out in South Africa while E.J. was responsible for analyses (HPLC) carried out in Sweden.

Conflicts of Interest: The authors declare no conflict of interest.

References

1. Tilahun, D.; Shiferaw, E.; Johansson, E.; Hailu, F. Genetic variability of Ethiopian bread wheat genotypes (*Triticum aestivum* L.) using agro-morphological traits and their gliadin content. *Afr. J. Agric. Res.* **2016**, *11*, 330–339.

2. Husenov, B.; Makhkamov, M.; Garkava-Gustavsson, L.; Muminjanov, H.; Johansson, E. Breeding for wheat quality to assure food security of a staple crop: The case study of Tajikistan. *Agric. Food Secur.* **2015**, *4*, 9. [CrossRef]

3. Zielinski, H.; Kozlowska, H. Antioxidant activity and total phenolics in selected cereal grains and their different morphological fractions. *J. Agric. Food Chem.* **2000**, *48*, 2008–2016. [CrossRef] [PubMed]

4. Hussain, A.; Larsson, H.; Kuktaite, R.; Johansson, E. Mineral composition of organically grown wheat genotypes: contribution to daily merals intake. *Int. J. Environ. Res. Public Health* **2010**, *7*, 3442–3456. [CrossRef] [PubMed]

5. Hussain, A.; Larsson, H.; Kuktaite, R.; Olsson, M.E.; Johansson, E. Carotenoid content in organically produced wheat: Relevance for human nutritional health on consumption. *Int. J. Environ. Res. Public Health* **2015**, *12*, 14068–14083. [CrossRef]

6. Johansson, E.; Hussain, A.; Kuktaite, R.; Andersson, S.C.; Olsson, M. Contribution of organically grown crops to human health. *Int. J. Environ. Res. Public Health* **2014**, *11*, 3870–3898. [CrossRef] [PubMed]

7. European Food Safety Authority. Scientific opinion on dietary reference values for vitamin E as α-tocopherol. EFSA Panel on dietetic products, nutrition and allergies (NDA). *EFSA J.* **2015**, *13*, 4149.

8. Wolf, G. The discovery of the antioxidant function of vitamin E: The contribution of Henry A. Mattill. *J. Nutr.* **2005**, *135*, 363–366. [PubMed]

9. Dörmann, P. Functional diversity of tocochromanols in plants. *Planta* **2007**, *225*, 269–276.

10. United States Department of Agriculture. USDA National Nutrient Database for Standard Reference. Available online: https://ndb.nal.usda.gov/ndb/ (accessed on 1 August 2017).

11. Hussain, A.; Larsson, H.; Olsson, M.E.; Kuktaite, R.; Grausgruber, H.; Johansson, E. Is organically produced wheat a source of tocopherols and tocotrienols for health food? *Food Chem.* **2012**, *132*, 1789–1795. [CrossRef]

12. Dror, D.K.; Allen, L.H. Vitamin E deficiency in developing countries. *Food Nutr. Bull.* **2011**, *2*, 124–143. [CrossRef] [PubMed]

13. Piironen, V.; Lampi, A.; Ekholm, P.; Salmenkallio-Marttila, M.; Liukkonen, K. Micronutrients and phytochemicals in wheat grain. In *Wheat Chemistry and Technology*, 4th ed.; Khalil, K.K., Shewry, P.R., Eds.; American Association of Cereal Chemists, Inc.: Eagan, MN, USA, 2009; pp. 179–210.

14. Lampi, A.; Nurmi, T.; Piironen, V. Efects of the environment and genotype on tocopherols and tocotrienols in wheat in the HEALTHGRAIN diversity screen. *J. Agric. Food Chem.* **2010**, *58*, 9306–9313. [CrossRef] [PubMed]

15. Wrigley, C.W. Wheat: A unique grain for the world. In *Wheat Chemistry and Technology*, 4th ed.; Khalil, K.K., Shewry, P.R., Eds.; American Association of Cereal Chemists, Inc.: Eagan, MN, USA, 2009; pp. 1–15.

16. Bramley, P.M.; Elmadfa, I.; Kafatos, A.; Kelly, F.J.; Manios, Y.; Roxborough, H.E.; Schuch, W.; Sheehy, P.J.A.; Wagner, K.H. Vitamin E. *J. Sci. Food Agric.* **2000**, *80*, 913–938. [CrossRef]

17. Fratianni, A.; Caboni, M.F.; Irano, M.; Panfili, G. A critical comparison between traditional methods and supercritical carbon dioxide extraction for the determination of tocochromanols in cereals. *Eur. Food Res. Technol.* **2002**, *215*, 353–358. [CrossRef]

18. Labuschagne, M.T.; Mkhatywa, N.; Wentzel, B.; Johansson, E.; Van Biljon, A. Tocochromanol concentration, protein composition and baking quality of white flour of South African wheat cultivars. *J. Food Compos. Anal.* **2014**, *33*, 127–131. [CrossRef]

19. Panfili, G.; Fratianni, A.; Irano, M. Normal phase-high performance liquid chromatography method for the determination of tocopherols and tocotrienols in cereals. *J. Agric. Food Chem.* **2003**, *51*, 3940–3944. [CrossRef] [PubMed]

20. Malik, A.H.; Kuktaite, R.; Johansson, E. Combined effect of genetic and environmental factors on the accumulation of proteins in the wheat grain and their relationships to bread-making quality. *J. Cereal Sci.* **2013**, *57*, 170–174. [CrossRef]

21. Moreira-Ascarrunz, S.D.; Larsson, H.; Prieto-Linde, M.L.; Johansson, E. Mineral nutritional yield and nutrient density of locally adapted wheat genotypes under organic production. *Foods* **2016**, *5*, 89. [CrossRef] [PubMed]

22. European Parliament and Council. Regulation (EU) No 1169/2011 of the European Parliament and of the Council of 25 October 2011 on the Provision of Food Information to Consumers. *Off. J. Eur. Commun.* **2011**, *L304*, 18.

23. Hidalgo, A.; Brandolino, A. Tocol stability during bread, biscuit and pasta processing from wheat. *J. Cereal Sci.* **2010**, *52*, 254–259. [CrossRef]

24. Zielinski, H.; Ciska, E.; Kozlowska, H. The cereal grains: Focus on vitamin E. *Czech J. Food Sci.* **2001**, *19*, 182–188.

25. Nurit, E.; Lyan, B.; Pujos-Guillot, E.; Branlard, G.; Piquet, A. Change in B and E vitamin and lutein, β-sitosterol contents in industrial milling fractiona and during toasted bread production. *J. Cereal Sci.* **2016**, *69*, 290–296. [CrossRef]

26. Andersson, S.C.; Olsson, M.E.; Gustavsson, K.-E.; Johansson, E.; Rumpunen, K. Tocopherols in rose hips (*Rosa* spp.) during ripening. *J. Sci. Food Agric.* **2012**, *92*, 2116–2121. [CrossRef] [PubMed]

27. Andersson, S.C.; Rumpunen, K.; Johansson, E.; Olsson, M.E. Tocopherols and tocotrienols in Sea Buchthorn (*Hoppophae rhamnoides*) berries during ripening. *J. Agric. Food Chem.* **2008**, *56*, 6701–6706. [CrossRef] [PubMed]

28. Delgado-Zamarreno, M.M.; Bustamante-Rangel, M.; Sierra-Manzano, S.; Verdugo-Jara, M.; Carabias-Martinez, R. Simultaneous extraction of tocotrienols and tocopherols from cereals using pressurized liquid extraction prior to LC determination. *J. Sep. Sci.* **2009**, *32*, 1430–1436. [CrossRef] [PubMed]

29. Sramkova, Z.; Gregova, A.; Sturdik, E. Chemical composition and nutritional quality of wheat grain. *Acta Chim. Slovaca* **2009**, *2*, 115–138.

30. Tiwari, U.; Cummins, E. Nutritional importance and effect of processing on tocols in cereals. *Trends Food Sci. Technol.* **2009**, *20*, 511–520. [CrossRef]

31. Serbinova, E.; Kagan, V.; Han, D.; Packer, L. Free-radical recycling and intramembrane mobility in the antioxidant properties of alpha-tocopherol and alpha-tocotrienol. *Free Radic. Biol. Med.* **1991**, *10*, 263–275. [CrossRef]

32. Schaffer, S.; Müller, W.E.; Eckert, G.P. Tocotrienols: Constitutional effects in aging and disease. *J. Nutr.* **2005**, *135*, 151–154. [PubMed]

33. Shibata, A.; Kobayashi, T.; Asai, A.; Eitsuka, T.; Oikawa, S.; Miyazawa, T.; Nakagawa, K. High purity tocotrienols attenuate atherosclerotic lesion formation in apoE-KO mice. *J. Nutr. Biochem.* **2017**, *48*, 44–50. [CrossRef] [PubMed]

34. Alawin, O.A.; Ahmed, R.A.; Dronamraju, V.; Briski, K.; Sylvester, P.W. γ-Tocotrienol-insduced disruption of lipid rafts in human breast cancer cells in associated with a reduction in exosome heregulin content. *J. Nutr. Biochem.* **2017**, *48*, 83–93. [CrossRef] [PubMed]

35. Allen, L.; Ramlingam, L.; Menikdiwela, K.; Scoggin, S.; Shen, C.-L.; Tomison, M.D.; Kaur, G.; Dufour, J.M.; Chung, E.; Kalupahana, N.S.; et al. Effects of delta-tocotrienol on obesity-related adipocyte hypertrophy, inflammation and hepatic steatosis in high-fat-fed mice. *J. Nutr. Biochem.* **2017**, *48*, 128–137. [CrossRef] [PubMed]

36. Fu, J.-Y.; Che, H.-L.; Tan, D.M.-Y.; Teng, K.-T. Bioavailability to tocotrienols: Evidence in human studies. *Nutr. Metab.* **2014**, *11*, 5. [CrossRef] [PubMed]

37. Miyazawa, T.; Shibata, A.; Sookwong, P.; Kawakami, Y.; Eitsuka, T.; Asai, A.; Oikawa, S.; Nakagawa, K. Antiangiogenic and anticancer potential of unsaturated vitamin E (tocotrienol). *J. Nutr. Biochem.* **2009**, *20*, 79–86. [CrossRef] [PubMed]

38. Sen, C.K.; Khanna, S.; Roy, S.; Packer, L. Molecular basis of vitamin E action—Tocotrienol potently inhibits glutamate-induced pp60(c-Src) kinase activation and death of HT4 neuronal cells. *J. Biol. Chem.* **2000**, *275*, 13049–13055. [CrossRef] [PubMed]

39. Tonini, T.; Rossi, F.; Claudio, P.P. Molecular basis of angiogenesis and cancer. *Oncogene* **2003**, *22*, 6549–6556. [CrossRef] [PubMed]

40. Miscellaneous Nutrients. Available online: https://www.dcnutrition.com/miscellaneous-nutrients/tocotrienols/ (accessed on 1 August 2017).

41. Lampi, A.; Nurmi, T.; Ollilainen, V.; Piironen, V. Tocopherols and tocotrienols in wheat genotypes in the HEALTHGRAIN diversity screen. *J. Agric. Food Chem.* **2008**, *56*, 9716–9721. [CrossRef] [PubMed]

42. Okarter, N.; Liu, C.; Sorrells, M.; Liu, R.H. Phytaochemical content and antioxidant activity of six diverse varieties of whole wheat. *Food Chem.* **2010**, *119*, 249–257. [CrossRef]

43. Wong, R.S.; Radhakrishnan, A.K. Tocotrienol research: Past into present. *Nutr. Res.* **2012**, *70*, 483–490. [CrossRef] [PubMed]

44. Fardet, A.; Rock, E.; Remesy, C. Is the in vitro antioxidant potential of whole-grain cereals and cereal products well reflected in vivo? *J. Cereal Sci.* **2008**, *48*, 258–276. [CrossRef]

45. Shewry, P.R.; Piironen, V.; Lampi, A.-M.; Edelmann, M.; Kariluoto, S.; Nurmi, T.; Fernandez-Orozco, R.; Ravel, C.; Charmet, G.; Andersson, A.A.M.; et al. The HEALTHGRAIN wheat diversity screen: Effects of genotype and environment on phytochemicals and dietary fiber components. *J. Agric. Food Chem.* **2010**, *58*, 9291–9298. [CrossRef] [PubMed]

46. Miller, H.E.; Rigelhof, F.; Marquart, L.; Prakash, A.; Kanter, M. Antioxidant content of whole grain beakfast cereals, fruits and vegetables. *J. Am. Coll. Nutr.* **2000**, *19*, 312–319. [CrossRef]

47. Baublis, A.J.; Lu, C.; Clydesdale, F.M.; Decker, E.A. Potential of wheat-based breakfast cereals as a source of dietary antioxidants. *J. Am. Coll. Nutr.* **2000**, *19*, 308–311. [CrossRef]

48. Perez-Jimenez, J.; Saura-Calixto, F. Literature data may underestimate the actual antioxidant capacity of cereals. *J. Agric. Food Chem.* **2005**, *53*, 5036–5040. [CrossRef] [PubMed]

49. Hidalgo, A.; Brandolino, A. Protein, ash, lutein and tocols distribution in einkorn (*Triticum monococcum* L. subsp. *monococcum*) seed fractions. *Food Chem.* **2008**, *107*, 444–448. [CrossRef]

Article

Lutein Esterification in Wheat Flour Increases the Carotenoid Retention and Is Induced by Storage Temperatures

Elena Mellado-Ortega and Dámaso Hornero-Méndez *

Chemistry and Biochemistry of Pigments Group, Food Phytochemistry Department, Instituto de la Grasa (CSIC), Campus Universidad Pablo de Olavide, Ctra. de Utrera km. 1, 41013 Seville, Spain; melladoortegae@gmail.com
* Correspondence: hornero@ig.csic.es; Tel.: +34-954-611550

Received: 9 November 2017; Accepted: 6 December 2017; Published: 11 December 2017

Abstract: The present study aimed to evaluate the effects of long-term storage on the carotenoid pigments present in whole-grain flours prepared from durum wheat and tritordeum. As expected, higher storage temperatures showed a catabolic effect, which was very marked for free carotenoid pigments. Surprisingly, for both cereal genotypes, the thermal conditions favoured the synthesis of lutein esters, leading to an enhanced stability, slower degradation, and, subsequently, a greater carotenoid retention. The putative involvement of lipase enzymes in lutein esterification in flours is discussed, particularly regarding the preferential esterification of the hydroxyl group with linoleic acid at the 3' in the ε-ring of the lutein molecule. The negative effects of processing on carotenoid retention were less pronounced in durum wheat flours, which could be due to an increased esterifying activity (the de novo formation of diesterified xanthophylls was observed). Moreover, clear differences were observed for tritordeum depending on whether the lutein was in a free or esterified state. For instance, lutein-3'-O-monolinoleate showed a three-fold lower degradation rate than free lutein at 37 °C. In view of our results, we advise that the biofortification research aimed at increasing the carotenoid contents in cereals should be based on the selection of varieties with an enhanced content of esterified xanthophylls.

Keywords: carotenoids; lutein esters; tritordeum; *Triticum turgidum* conv. *durum*; carotenoid retention; whole-grain flour

1. Introduction

Carotenoids, the most widespread pigments in nature, are liposoluble antioxidants produced by plants, algae, fungi, and some bacteria. In plants, carotenoids contribute to the photosynthetic process by acting as light collectors and photoprotectors [1]. Moreover, carotenoids are found in high concentrations in most fruits and flowers where they contribute to the bright colours that attract animals for seed and pollen dispersion. These ubiquitous pigments can also be found in some roots, tubers, and grains, mostly due to the selection of coloured varieties as desired traits for plant domestication by man [2]. Animals are not able to synthetize carotenoids de novo; consequently carotenoids must be acquired through dietary consumption.

Different carotenoids are associated with different health benefits. The provitamin A activity of carotenoids with at least one unsubstituted β-ring end-group in their structure is well-known and of nutritional significance [3]. Moreover, epidemiological studies have correlated carotenoid intake with protection against a range of chronic diseases, such as cardiovascular diseases and cancer [4,5]. In particular, lutein and zeaxanthin play important roles in the prevention of eye diseases such as age related macular degeneration (AMD), cataracts, and retinitis pigmentosa [6].

Wheat (*Triticum* spp.), one of the most important crops for human consumption worldwide, contains low carotenoid contents compared to most vegetables and fruits. However, the widespread and daily-based consumption of cereals and derived products makes these staple foods an important source of these antioxidants in the diet, particularly in disadvantaged populations. The main objective of biofortification programs is the breeding of crops for better nutrition. The major carotenoid present in wheat is lutein [7]. Among the different *Triticum* species, durum wheat (*Triticum turgidum* ssp. *durum*) grains are characterized for presenting a yellowish colour due to carotenoids. In fact the yellow colour of pasta is a major quality trait [8]. Manipulating the carotenoid content of several cereals, or other crops, by means of biofortification strategies has the potential to provide significant health benefits without altering normal dietetic habits [9]. Remarkable progress has been made in this area; a good example is HarvestPlus, part of the Consultative Group on International Agriculture Research (CGIAR) Program on Agriculture for Nutrition and Health (A4NH). HarvestPlus collaborators have developed staple crops with increased densities of micronutrients through plant breeding techniques, such as pro-vitamin A cassava, which provides up to 40% of the daily vitamin A requirement, iron pearl millet, which provides up to 80% of daily iron needs, and zinc rice, which provides up to 60% of the daily zinc requirement (http://www.harvestplus.org/).

The *Triticeae* tribe, which includes wheat, barley (*Hordeum vulgare*), and rye (*Secale cereale*) species, is a series of closely related polyploids. Fertile amphiploid hybrids can be generated among the different cultivated members of this tribe and some wild cereal species. Tritordeum (*Tritordeum*; $2n = 6x = 42$, AABBH^{ch}H^{ch}) is a cereal obtained from the cross between a wild barley (*Hordeum chilense* Roem. & Schult.) with diploid genome (H^{ch}H^{ch}) and durum wheat (*Triticum turgidum* conv. *durum*; [10]). The lutein content in tritordeum is about 5–8 times higher than durum wheat and is characterized by a specific esterification profile involving two major fatty acids (linoleic and palmitic acids) [11,12]. The latter characteristic is derived from the genetic background of *H. chilense* [13]. The detailed composition of the lutein esters present in tritordeum has been determined and consists of four monoesters (lutein 3'-O-linoleate, lutein 3-O-linoleate, lutein 3'-O-palmitate, and lutein 3-O-palmitate) and four diesters (lutein dilinoleate, lutein 3'-O-linoleate-3-O-palmitate, lutein 3'-O-palmitate-3-O-linoleate, and lutein dipalmitate) [12]. Tritordeum is currently the subject of an intense breeding program at the Institute of Sustainable Agriculture (IAS; http://www.ias.csic.es/en/) in Cordoba, Spain, to optimize its use as a cereal for incorporation into the formulation of both functional and novel foods.

Cereal grains are traditionally processed for human consumption. The influence of processing techniques on the composition of phytonutrients and bioactive compounds of many staple foods, including cereal grains, has been extensively studied. For example, in cassava, maize, and sweet potato, the conditions and duration of storage have a more significant negative impact on the retention of provitamin A carotenoids than drying or cooking [14]. Similarly, Mugode et al. [15] concluded that the degradation of provitamin A carotenoids in maize mostly occurred during storage and this effect varied among genotypes. Whole-grain wheat flour contains substantially more vitamins, minerals, antioxidants, and other nutrients, including carotenoids, than refined wheat flour [16]. In addition to milling, the subsequent storage of grains and flours can have a significant impact on the composition of phytochemicals. In fact, cereal flour is usually stored during prolonged periods as part of their industrial and technological treatments, and, therefore, an important impact on the carotenoid content is expected as a result of storage. In the case of wheat flour, the main cause of carotenoid degradation is oxidation (including both enzymatic and non-enzymatic processes) during the storage period. Oxygen present in the medium is considered to be the major factor affecting the stability of carotenoids [17]. Therefore, in addition to considering the "high pigment content" trait of *Triticeae* genotypes, the "carotenoid retention ability" should also be considered when screening and selecting strains for their inclusion in breeding programs [18]. Moreover, new technological treatments are emerging for the processing of cereals and their derived products in order to preserve and enhance the content of carotenoids and other phytonutrients of nutritional relevance (reviewed by Hemery et al. [19]).

The natural process by which xanthophylls are esterified with fatty acids is an important part of post-carotenogenic metabolism and mediates their accumulation in plants. To assist in the development of carotenoid-enhanced cereals, the biochemical characterization of the xanthophyll esterification process and studies of the capacity of the cereal endosperm tissue to store these pigments are necessary. Some studies have highlighted the importance of the carotenoid retention capacity, and its influence on the stability during the postharvest storage of crops [20,21]. However, only a few studies investigated the involvement of xanthophyll esterification in the retention of carotenoids during the storage of cereals and derived products [22,23].

We recently assessed the effect of long-term storage on the biosynthesis of lutein esters in durum wheat and tritordeum grains and found that xanthophyll esterification was induced by environmental conditions (especially the temperature) [24]. We also found the xanthophyll esterification process to be highly specific (with the preferential esterification of lutein at position 3 of the β-end ring) and that the fatty acids involved in the esterification and their position in the lutein molecule had a significant effect on the carotenoid stability. Although the results from our previous study increased our understanding of the effects of long-term storage on carotenoid metabolism, characterization of the xanthophyll esterification process was lacking. Therefore, the main goal of the present study was to fully categorize the stability of carotenoid pigments in cereal flours and the influence of pigment esterification during the long-term storage of whole-grain flours in different temperature-controlled conditions.

2. Materials and Methods

2.1. Plant Material, Sample Preparation and Storage Conditions

A commercial durum wheat variety (Don Pedro) and a high-carotenoid tritordeum line (HT621, germplasm line developed in the framework of the Cereal Breeding Program carried out at the Institute for Sustainable Agriculture, Córdoba, Spain) [25] were used in the present study. Both samples are considered representatives of these two cereal genotypes and have been previously characterized regarding their carotenoid profile [11,12,23,26]. Plants were grown in 1-L pots, until maturity, under greenhouse conditions with supplementary lights providing a day/night regime of 12/12 h at 22/16 °C. Immediately after harvesting, seeds were preserved for 2 months at 4 °C in a desiccator before the beginning of experiment. After this storage period, and for each cereal genotype, whole-grain flour was obtained from 500 g of grains by using an oscillating ball mill Retsch Model MM400 (Retsch, Haan, Germany) at 25 Hz for 1 min. Subsequently, the resulting flour was distributed in lots of approximately 4 g in round-capped polypropylene 15-mL centrifuge tubes. Flour samples were stored under controlled temperature conditions (−32, 6, 20, 37 and 50 °C) for a period of 12 months. A control sample (t = 0 days) consisting of 5 subsamples was taken and analysed for each cereal type. Triplicate samples (three tubes for each temperature and time) were taken at monthly intervals and analysed in duplicate. The dry matter content (%) in the samples at each sampling date was measured in triplicate by using an Ohaus moisture balance model MB35 (Ohaus, Greifensee, Switzerland). During the course of the experience a continuous monitoring of the storage temperature was performed.

2.2. Chemicals and Reagents

Deionised water (HPLC-grade) was produced with a Milli-Q Advantage A10 system (Merck Millipore, Madrid, Spain). HPLC-grade acetone was supplied by BDH Prolabo (VWR International Eurolab, S.L., Barcelona, Spain). The rest of reagents were all of analytical grade.

2.3. Extraction of Carotenoids

Carotenoid pigments were extracted from flours with the following procedure. Briefly, 1 g of flour was placed into 25 mL stainless-steel grinding jar together with two stainless-steel balls (15 mm ø), 6 mL of acetone containing 0.1% (w/v) BHT and a known amount of internal standard (β-apo-8′-carotenal; 1.75 and 3.50 μg for durum wheat and tritordeum samples, respectively). Samples were crushed in an oscillating ball mill Retsch Model MM400 (Retsch, Haan, Germany) at 25 Hz for 1 min. Most of the resulting slurry was transferred into a tube and centrifuged at $4500 \times g$ for 5 min at 4 °C and the clear supernatant collected in a clean tube. The solvent was gently evaporated under a nitrogen stream, and the pigments were dissolved in 0.5 mL of acetone. Prior to the chromatographic analysis, samples were centrifuged at $13,000 \times g$ for 5 min at 4 °C. The analyses were carried out in duplicate for each sample. All operations were performed under dimmed light to prevent isomerization and photo-degradation of carotenoids.

2.4. HPLC Analysis of Carotenoids

The procedures for the isolation and identification of carotenoid pigments and its esters have already been described in previous works [11,12]. Quantitative analysis of carotenoids was carried out by HPLC according to Atienza et al. [11]. The HPLC system consisted of a Waters e2695 Alliance chromatograph fitted with a Waters 2998 photodiode array detector, and controlled with Empower2 software (Waters Cromatografía, S.A., Barcelona, Spain). A reversed-phase column (Mediterranea SEA18, 3 μm, 20 × 0.46 cm; Teknokroma, Barcelona, Spain) was used. Separation was achieved by a binary-gradient elution using an initial composition of 75% acetone and 25% deionized water, which was increased linearly to 95% acetone in 10 min, then raised to 100% in 2 min, and maintained constant for 10 min. Initial conditions were reached in 5 min. An injection volume of 10 μL and a flow rate of 1 mL/min were used. Detection was performed at 450 nm, and the UV-visible spectra were acquired online (350–700 nm wavelength range). Quantification was carried out using calibration curves prepared with lutein, zeaxanthin and β-carotene standards isolated and purified from natural sources [27]. Calibration curves including eight-points were prepared in the pigment concentration range of 0.5–45 μg/mL. Lutein ester content were estimated by using the calibration curve for free lutein, since the esterification of xanthophylls with fatty acids does not modify the chromophore properties [28]. Accordingly, the concentration of lutein esters was expressed as free lutein equivalents. The calibration curve of free lutein was also used to determine the concentration of the (Z)-isomers of lutein. Data were expressed as μg/g dry weight (μg/g dw).

2.5. Degradation Kinetics Model

In order to investigate the effects of the esterification on the carotenoid degradation during the storage period, the reaction order and derived kinetic parameters were only investigated in those time ranges where the occurrence of catabolic reactions was dominant for each pigment (that is a decline in the concentration with time was clearly observed). For this purpose zero- and first-order kinetics were hypothesized. The general reaction rate expression was applied, $-dC/dt = kC^n$, where C is the concentration of the compound (μg/g dw), k is the reaction rate constant (months^{-1}), t is the reaction time (months), and n is the order of the reaction [29]. The reaction rate expression and the kinetic parameters for zero- and first-order models are summarized in Table 1. The selected order of the reaction was that showing the best correlation (R^2) and the best correspondence among the experimental values and the half-life of the compound ($t_{1/2}$) and D ($t_{1/10}$) [time needed for the concentration of a reactant to fall to half and one tenth its initial value respectively, where $t_{1/2} = C_0/2k$ and $t_{1/10} = 0.9C_0/k$ for zero-order and $t_{1/2} = (Ln2)/k$ and $t_{1/10} = (Ln10)/k$ for first-order]. Kinetic parameters derived from fitted models with $R^2 < 0.8$ were not considered in the discussion.

Table 1. Expression of the reaction rate depending on the reaction order (*n*) and kinetic parameters derived.

Reaction Order	Reaction Rate Expression	Integrated Expression	Graphical Representation	Half-Life [a] $(t_{1/2})$	D [b] $(t_{1/10})$
Zero *n* = 0	$-dC/dt = kC^0 = k$	$C-C_0 = -kt$	$C-C_0$ vs. t Slope $= -k$	$t_{1/2} = C_0/2k$	$t_{1/10} = 0.9C_0/k$
First *n* = 1	$-dC/dt = kC^1 = kC$	$Ln(C/C_0) = -kt$	$Ln(C/C_0)$ vs. t Slope $= -k$	$t_{1/2} = Ln(2)/k$	$t_{1/10} = Ln(10)/k$

[a] Time needed for the concentration of a reactant to fall to half of its initial value. [b] Time needed for the concentration of a reactant to fall one tenth of its initial value.

2.6. Statistical Analysis

Pigment contents are expressed as mean and standard error of the mean (SEM). Significant differences between means was determined by one-way ANOVA, followed by a post-hoc test of mean comparison using the Duncan test for a confidence level of 95% ($p < 0.05$) utilizing the STATISTICA 6.0 software (StatSoft Inc., Tulsa, OK, USA).

3. Results and Discussion

3.1. Carotenoid Content in Whole-Grain Flours: Effect of Long-Term Storage

The initial carotenoid composition for the flours of durum wheat (*Triticum turgidum* conv. *durum*, Don Pedro) and tritordeum (Tritordeum HT621 line) was consistent with previous studies (see Table 2 at t = 0 months) [11,12,23,26]. The initial concentrations of individual pigments were higher in whole-grain tritordeum flour than durum wheat flour, with the exception of (all-*E*)-zeaxanthin, which is not present in tritordeum. On average, the total initial carotenoid content was six times higher in tritordeum (HT621 line) with respect to durum wheat (Don Pedro).

Table 2. Initial carotenoid composition in *Triticum turgidum* cv. *durum* (Don Pedro variety) and *Tritordeum* (HT621 line) whole-grain flours subjected to long-term storage (12 months) under controlled temperature.

HPLC Peak [a]	Pigment	Concentration (µg/g Dry Weight) [b]	
		Durum Wheat (Don Pedro Variety)	Tritordeum (HT621 Advanced Line)
1	(all-*E*)-Zeaxanthin	0.08 ± 0.00	-
2	(all-*E*)-Lutein	1.08 ± 0.02	3.95 ± 0.04
3	(9*Z*)-Lutein	0.06 ± 0.00	0.19 ± 0.01
4	(13*Z*)-Lutein	0.12 ± 0.00	0.30 ± 0.01
5	Lutein-3'-*O*-linoleate	0.01 ± 0.00	0.15 ± 0.00
6	Lutein-3-*O*-linoleate	0.01 ± 0.00	0.72 ± 0.01
5 + 6	Lutein monolinoleate	0.03 ± 0.00	0.87 ± 0.01
7	Lutein-3'-*O*-palmitate	0.00 ± 0.00	0.49 ± 0.01
8	Lutein-3-*O*-palmitate	0.00 ± 0.00	1.01 ± 0.02
7 + 8	Lutein monopalmitate	0.01 ± 0.00	1.50 ± 0.01
9	(all-*E*)-β-Carotene	0.02 ± 0.00	0.06 ± 0.00
10	Lutein-3,3'-dilinoleate	n.d. [c]	0.12 ± 0.00
11	Lutein-3'-*O*-linoleate-3-*O*-palmitate plus Lutein-3'-*O*-palmitate-3-*O*-linoleate	n.d.	0.42 ± 0.01
12	Lutein-3,3'-dipalmitate	n.d.	0.41 ± 0.01
	Lutein monoesters	0.04 ± 0.00	2.37 ± 0.01
	Lutein diesters	-	0.96 ± 0.01
	Total lutein esters	0.04 ± 0.00	3.33 ± 0.02
	Total free lutein	1.26 ± 0.02	4.44 ± 0.09
	Total lutein	1.29 ± 0.02	7.77 ± 0.07
	Total carotenoids	1.39 ± 0.03	7.83 ± 0.07
	Regioisomers ratios		
	Lutein-3-*O*-linoleate/Lutein-3'-*O*-linoleate	1	5
	Lutein-3-*O*-palmitate/Lutein-3'-*O*-palmitate	1	2

[a] Peak numbers according to Figure S1 (Supplementary Materials). [b] Data represent the mean ± standard error (*n* = 5). [c] n.d. not detected.

The evolution of total carotenoids revealed significant losses for both cereals at the higher examined temperatures of 37 and 50 °C (Figure 1; see also Figure S1, and Tables S1 and S2). The total degradation of pigments was observed by the end of the 12-month storage period in the durum wheat and tritordeum samples kept at the higher temperature (50 °C). In accordance with Gayen et al. [30], the experimental conditions tested in the present study may facilitate the action of degradative enzymes, such as lipoxygenase (LOX), leading to the co-oxidation of carotenoid pigments. LOX is mostly located in the germ and bran of the grain kernel and its main substrate is linoleic acid [31]. In addition to LOX degradation, the susceptibility of carotenoids to oxygen and high temperatures should be responsible for the decrease in the pigment levels observed during our experiments. The pigment content reduction at the end of the 12-month storage period at −32, 6, and 20 °C were similar in all cases for both cereal genotypes. In contrast, significant differences were observed between the two cereals at 37 °C, with a greater retention of pigments observed in tritordeum (pigment losses of 84% for durum wheat compared to 72% for tritordeum) (Figure 1).

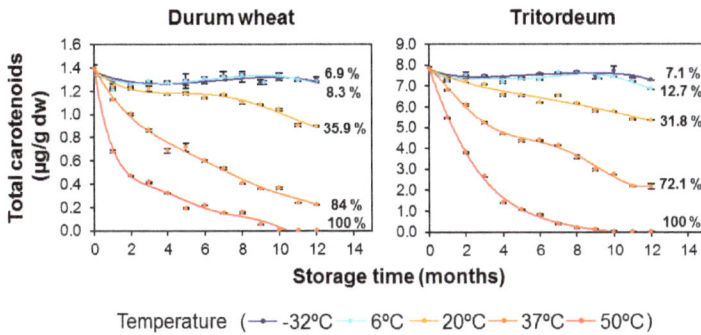

Figure 1. Evolution of the total carotenoid content (μg/g dry weight) in durum wheat (Don Pedro variety) and tritordeum (HT621 advanced line) whole-grain flours during long-term storage under temperature controlled conditions (−32, 6, 20, 37 and 50 °C). The values shown are the mean and standard error (*n* = 5 for the starting sample, *n* = 3 for the rest of the samples). Pigment losses (%) are indicated at the end of storage.

The changes observed for the individual free pigments in durum wheat (Figure 2) were similar to the trend described for total carotenoid content: The carotenoid content remained fairly constant at the lowest storage temperatures of −32 and 6 °C while higher carotenoid losses were observed with increased storage temperature. Thus, the carotenoid losses after 12 months of storage at 20 °C were 51% and 63% for (all-*E*)-lutein and (all-*E*)-zeaxanthin, respectively (Table S1). The declines were lower at 20 and 37 °C for the (*Z*)-isomers of lutein (39% and 89%, respectively), which is consistent with a *trans* to *cis* isomerization process, as reported in other similar studies [32]. At the end of the storage period at 37 °C, (all-*E*)-lutein, (all-*E*)-zeaxanthin, and (all-*E*)-β-carotene had been degraded to trace levels. For the durum wheat flour maintained at 50 °C, (all-*E*)-zeaxanthin and (all-*E*)-β-carotene were already undetectable at the seventh month. These results suggest that both zeaxanthin and β-carotene are less thermostable than lutein at higher temperatures. Some authors have suggested that (all-*E*)-β-carotene is the most thermolabile carotenoid, especially in low-water environments as in the case of cereal flour [33].

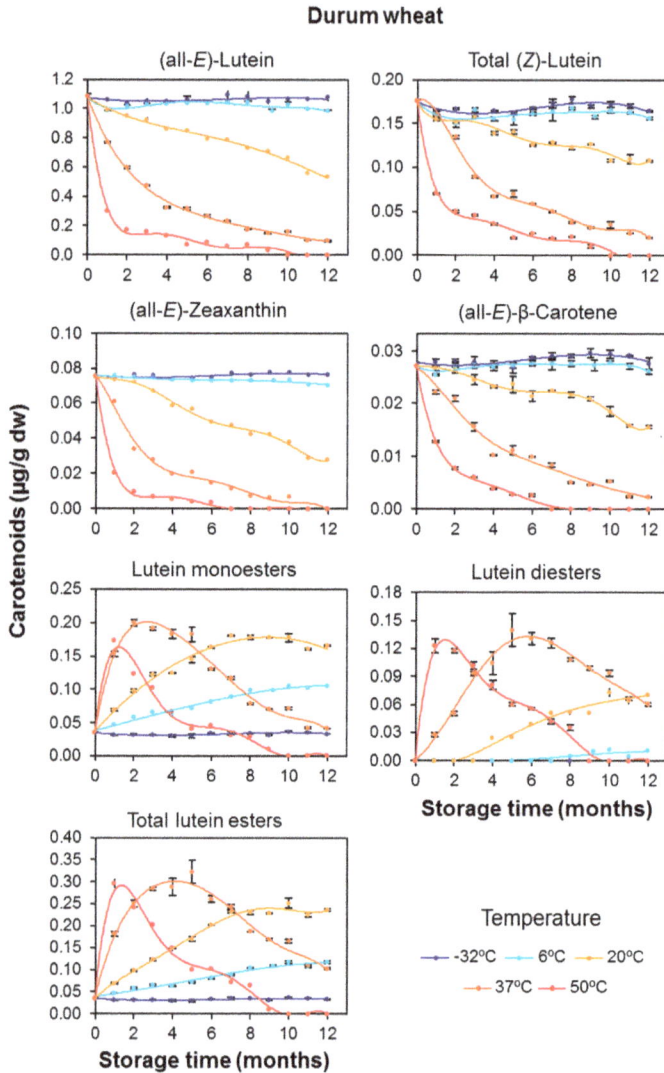

Figure 2. Evolution of the individual carotenoid content and esterified fractions (µg/g dry weight) in durum wheat (Don Pedro variety) whole-grain flours during long-term storage under temperature controlled conditions (−32, 6, 20, 37 and 50 °C). The values shown are the mean and standard error (*n* = 5 for the starting sample, *n* = 3 for the rest of the samples).

For whole-grain tritordeum flour, greater losses of the individual free pigments were also observed at the higher storage temperatures (Figure 3). For tritordeum flour stored at 37 °C, (all-*E*)-lutein, (*Z*)-lutein, and (all-*E*)-β-carotene had decreased by 83–98% by the end of the 12-month storage period (Table S2). Similar to durum wheat, the total pigments in tritordeum were completely destroyed after 10 months of storage at 50 °C, and (all-*E*)-β-carotene was already undetectable after 4 months.

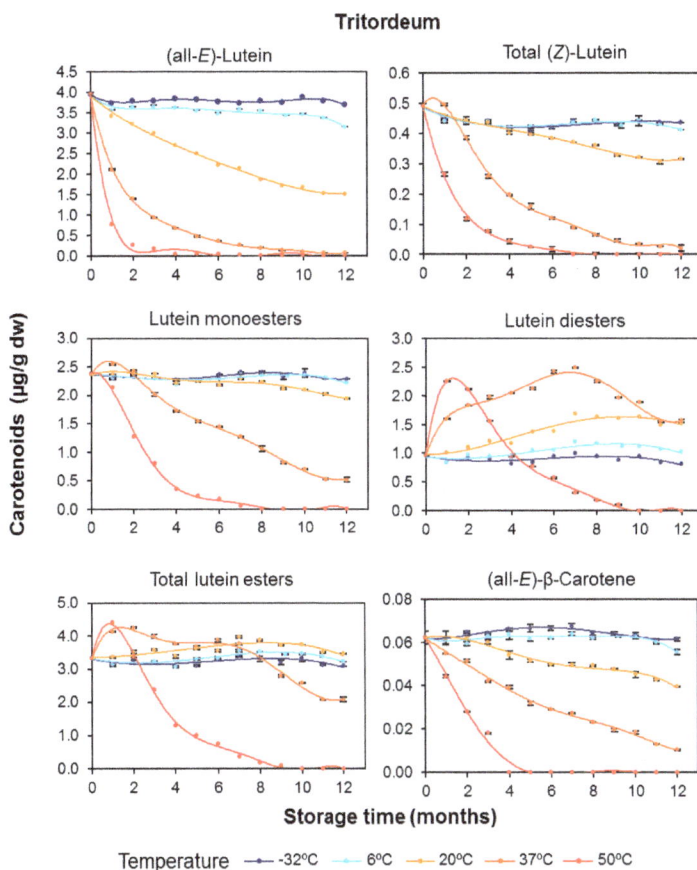

Figure 3. Evolution of the individual carotenoid content and esterified fractions (µg/g dry weight) in tritordeum (HT621 advanced line) whole-grain flours during long-term storage under temperature controlled conditions (−32, 6, 20, 37 and 50 °C). The values shown are the mean and standard error (*n* = 5 for the starting sample, *n* = 3 for the rest of the samples).

In contrast to the free carotenoids, the lutein esters showed increased levels with increasing temperature. This distinction was particularly striking for durum wheat, and especially the monoester fraction, stored at the milder conditions of 6 and 20 °C (Figure 2). For example, the concentration of lutein monoesters increased by 3.7 and 5.7 times the initial values after 12 months at 6 and 20 °C, respectively. At the higher temperatures (37 and 50 °C), competition and/or compensation was observed between the increases promoted by the temperature (de novo esterification) and decreases caused by oxidative degradation. After the storage of durum wheat for 12 months, concentration increases of 2–3 fold were recorded at 37 °C for the lutein ester fraction (sum of monoesters and diesters), revealing an intense esterifying activity under these conditions. The concomitant formation of lutein diesters associated with the rise in storage temperature and time was prompted by the increase in the pool of monoesters.

The free pigments in tritordeum showed a general pattern of degradation with increasing temperature; in contrast, analysis of the esterified fractions in tritordeum revealed clear differences in the evolution profile of lutein monoesters and diesters (Figure 3). Similar to the free pigments, the monoesters fraction decreased by 18.4% and 78.3% with respect to initial levels after 12-months storage at 20 and 37 °C, respectively. However, the diester fraction showed a similar behaviour to that found in durum wheat, with increases of 7%, 60%, and 63% with respect to initial levels after 12-months storage at 6, 20, and 37 °C, respectively. These results might indicate either the occurrence of different isoforms of the esterifying enzymes in the two cereals, or a more intensive esterifying activity of the enzymes in durum wheat. The *H. chilense* genetic background of tritordeum could be responsible for these differences.

The synthesis of lutein esters in cereal flours could possibly take place through a different metabolic pathway from the one operating in intact grains. In fact, in a preliminary study carried out with durum wheat flour [23], the formation of lutein esters was observed, especially at high temperatures; however, the trace levels of the pigments impaired their quantification. The present results underline the importance of controlling the storage conditions of cereals and derived products in order to prevent or promote changes in the profile of carotenoids and other phytochemicals. One possibility is that the esterification of xanthophylls in flours is mediated by the activity of lipases, which are concentrated in the bran in the case of cereals [34]. Lipases catalyse the hydrolysis of carboxyl-ester linkages leading to the release of fatty acids and organic alcohols. However, under low-water conditions, lipases may catalyse the reverse reaction (esterification) or various transesterification reactions involving acids, alcohols, and esters [35]. The acyl transferase activity of lipases has already been suggested to be involved in the formation of sterol esters during the long-term storage of wheat flour [36]. These authors also observed the esterification of lutein, although the data obtained were inconclusive. More recently, Ahmad et al. [37] have provided sound data concluding that lutein esterification in bread wheat was genetically controlled and likely due to a GDSL-lipase located on the 7D chromosome and co-located with a QTL (Quality Trait Loci) associated with ester formation. Moreover, Mattera et al. [38,39] demonstrated that lutein esterification in wheat endosperm is controlled by chromosomes of the homoeologous group 7 (7D and 7H^{ch}), and suggested differential fatty acid enzyme specificity depending on the cereal species (for instance, common wheat and *H. chilense*).

3.2. Effect of Long-Term Storage on the Esterified Lutein Fractions

The evolution of individual lutein monoesters (lutein monolinoleate and monopalmitate) was similar and consistent with the corresponding changes observed for the total esters fraction during the entire storage of both cereal flours. The profiles at 37 and 50 °C presented two distinctive areas of net synthesis (peaks) and degradation (troughs). This was particularly evident in durum wheat (Figure 4). At 37 °C, lutein monolinoleate showed a synthesis period over the first 2 months of storage, with a 4-fold concentration increase, followed by a degradation period during the remaining storage time, reaching up to 78% total degradation. Similarly, at 50 °C, the maximum synthesis was observed after the first month, accounting for concentrations of up to 3 times the initial content. Lutein monopalmitate showed a more pronounced synthesis, with a 10-fold increase observed after two months at 37 °C and an 8-fold increase after the first month at 50 °C (Figure 4). These data indicate differences in relation to the preferential formation of lutein monoesters with both fatty acids, with the palmitic acid esters being more abundant. On the other hand, the degradation rate was similar for lutein monopalmitate and lutein monolinoleate during the storage of durum wheat flour at 37 and 50 °C. Thus, the ratio between lutein monopalmitate and lutein monolinoleate remained constant across the whole storage period (Table S1). However, in a previous study with durum wheat grains submitted to long-term storage, a higher thermostability for lutein monolinoleate was reported [24].

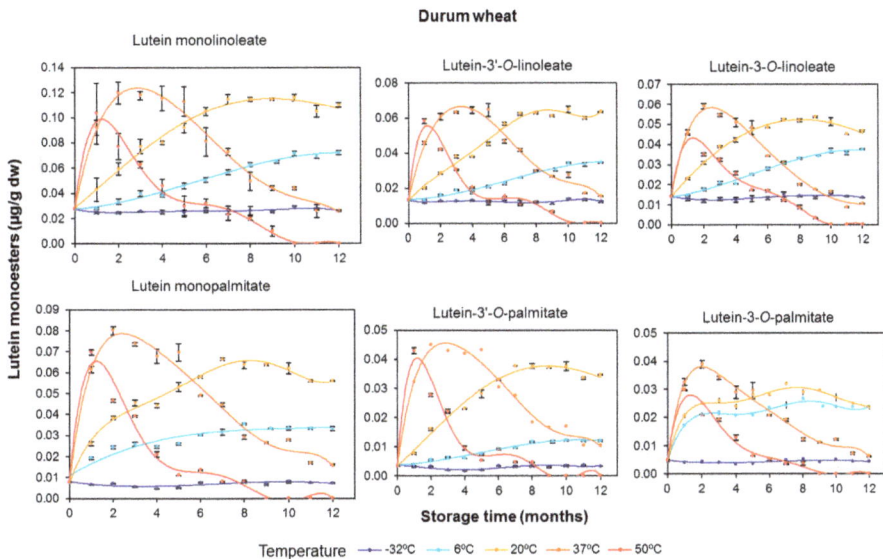

Figure 4. Quantitative changes in the xanthophyll ester fractions (μg/g dry weight) in durum wheat (Don Pedro variety) whole-grain flours during long-term storage under temperature controlled conditions (−32, 6, 20, 37 and 50 °C). The values shown are the mean and standard error (*n* = 5 for the starting sample, *n* = 3 for the rest of the samples).

In tritordeum, the maximum concentration increase for lutein monolinoleate was registered after two months at 37 °C and one month at 50 °C (Figure 5). In contrast to durum wheat, stability differences between both monoesters were observed in tritordeum. Lutein monolinoleate remained mostly constant at −32, 6, and 20 °C, whereas lutein monopalmitate showed a progressive decrease with increasing temperature and storage time. Nevertheless, the degradation rates for lutein monoesters were consistently lower than the rates recorded for free lutein (Figure 3). The degradation after 12 months at 37 °C was greater for lutein monopalmitate (85.9%) than for lutein monolinoleate (65.1%).

With respect to the increases in the lutein ester contents observed in durum wheat (Figures 2 and 4), the putative lipase enzyme could exhibit greater activity for palmitic acid than for linoleic acid due to a higher specificity for palmitic acid as a substrate or a better availability of this saturated fatty acid. These results are in line with those obtained by O'Connor et al. [40] who reported the lipase activity in different cereals. However, the evolution of the esterified fraction in tritordeum flour suggested a different scenario compared to durum wheat. In order to interpret this data, it should be taken into consideration that tritordeum is a novel hybrid cereal in which the *H. chilense* genome contributes and interacts with that of durum wheat. Therefore, it is likely that these genetic differences are the cause for the lower esterification activity in tritordeum flour.

Regarding the two regioisomers for each lutein monoester, the 3′ position of the lutein molecule appears to be the preferred site for esterification mediated by lipases in both cereals (Figures 4 and 5), and gives rise to a more stable compound (i.e., lutein-3′-O-linoleate and lutein-3′-O-palmitate are more stable than lutein-3-O-linoleate and lutein-3-O-palmitate, respectively). These observations reinforced the idea that the enzyme systems involved in the esterification of lutein in cereal flours are different from those operating in intact grains [12].

In both cereal flours, lutein diesters presented remarkable net increases during the duration of the storage period, especially for lutein dilinoleate and particularly at the storage temperature of 37 °C (Figures 6 and 7). The evolution of diesterified xanthophylls over the storage period clearly suggests that their formation is due to induction of the esterification process by temperature, with this synthesis being particularly prominent in durum wheat. As the storage temperature increased, the rate and amounts of synthesis were also increased, but, on the other hand, the degradative processes due to oxidative stress were also induced. Interestingly, the higher the degree of esterification (diester > monoester > free), the higher the xanthophyll's stability and, consequently, the more delayed the degradation (Figures 4 and 6 for durum wheat and Figures 5 and 7 for tritordeum).

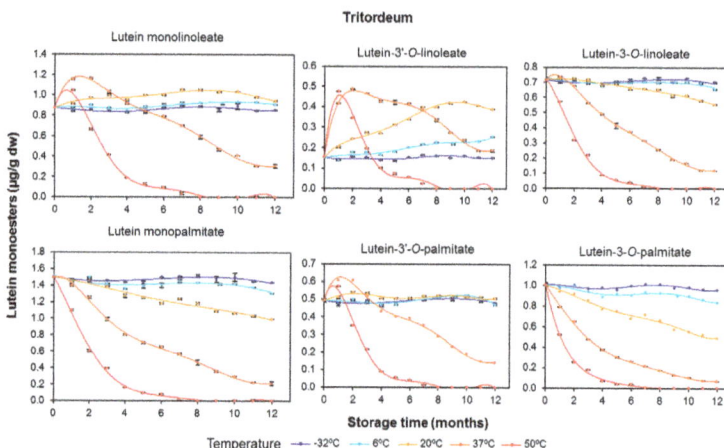

Figure 5. Evolution of the monoesterified lutein (μg/g dry weight), including the regioisomers, in tritordeum (HT621 advanced line) whole-grain flours during long-term storage under temperature controlled conditions (−32, 6, 20, 37 and 50 °C). The values shown are the mean and standard error ($n = 5$ for the starting sample, $n = 3$ for the rest of the samples).

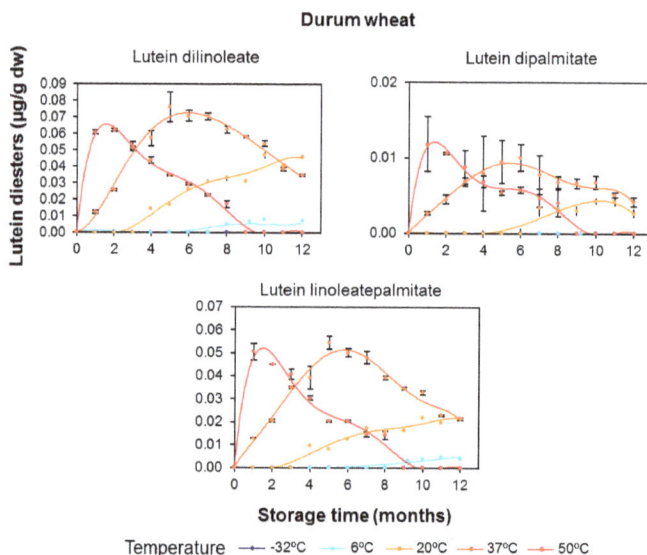

Figure 6. Evolution of the diesterified lutein (μg/g dry weight) in durum wheat (Don Pedro variety) whole-grain flours during long-term storage under temperature controlled conditions (−32, 6, 20, 37 and 50 °C). The values shown are the mean and standard error ($n = 5$ for the starting sample, $n = 3$ for the rest of the samples).

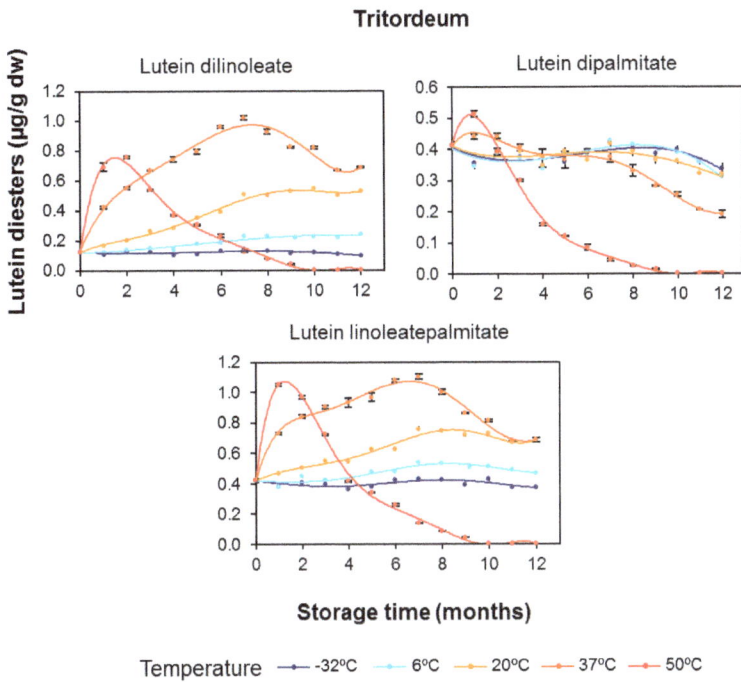

Figure 7. Evolution of the diesterified lutein (μg/g dry weight) in tritordeum (HT621 advanced line) whole-grain flours during long-term storage under temperature controlled conditions (−32, 6, 20, 37 and 50 °C). The values shown are the mean and standard error (*n* = 5 for the starting sample, *n* = 3 for the rest of the samples).

3.3. Kinetics of Retention of Carotenoids during the Long-Term Storage of Wheat Flours

A kinetic study indicated a progressive increase of the rate constants (*k*) with the rise of temperature for both cereal genotypes. Tables 3 and 4 summarized the kinetic data characterizing the evolution of both free and total pigments (including esterified pigments) during long-term storage assuming zero- and first-order kinetic models, respectively. As deducted from the correlation coefficient values, the first-order kinetic model showed best adjustment to the data, indicating that the degradation reaction rate is directly proportional to the pigment concentration ($-dC/dt = kC$; see Materials and Methods and Table 1). These results are consistent with those reported by other authors [32,41]. The reaction rates were higher in tritordeum for all pigments at 20, 37, and 50 °C, with β-carotene being an exception. At 50 °C, the *k* values for all xanthophylls were approximately double in tritordeum compared to durum wheat. Pigment structure (free or esterified with fatty acids), matrix effects (including the presence of oxidative enzymes and other antioxidants), and the oxidative stressing environment are likely to be key factors for this phenomena. The milling process, involving cell and tissue disruption, and the subsequent storage of cereal flours have also previously been found to affect carotenoid stability [42]. The possible presence of other antioxidants might also produce a protective effect on the carotenoids. For example, the changes and/or interactions between tocopherols and carotenoids during cereal processing have been analysed by several authors [43,44]. However, there is no information about such interactions in tritordeum, with further work needed in this area.

Table 3. Reaction rate constant (k; month^{-1}) for the total carotenoid content in durum wheat (Don Pedro variety) and tritordeum (HT621 line) whole-grain flours during a long-term storage period (12 months) at -32, 6, 20, 37, and 50 °C following the zero-order kinetic model ($C-C_0 = -kt$).

Pigment	T (°C)	Durum Wheat (Don Pedro Variety)		Tritordeum (HT621 Advanced Line)	
		k ($\times10^{-3}$ Month^{-1})	R^2	k ($\times10^{-3}$ Month^{-1})	R^2
(all-*E*)-Zeaxanthin	-32	0.2	0.33	-	-
	6	0.4	0.80	-	-
	20	4	0.98	-	-
	37	5	0.80	-	-
	50	3	0.43	-	-
(all-*E*)-Lutein	-32	1	0.06	7	0.15
	6	3	0.18	41	0.72
	20	42	0.98	198	0.96
	37	68	0.80	232	0.65
	50	51	0.47	160	0.33
Total free (*Z*)-Lutein	-32	0.3	0.04	2	0.14
	6	0.2	0.01	3	0.31
	20	5	0.92	15	0.97
	37	12	0.86	42	0.87
	50	10	0.63	28	0.57
(all-*E*)-β-Carotene	-32	0.1	0.35	0.1	0.02
	6	0.5	0.06	0.2	0.12
	20	8	0.93	2	0.95
	37	2	0.90	4	0.97
	50	1	0.62	4	0.62
Total free lutein	-32	1	0.06	9	0.19
	6	3	0.15	42	0.70
	20	47	0.98	213	0.95
	37	80	0.82	273	0.70
	50	60	0.50	188	0.36
Total free carotenoids	-32	1	0.01	9	0.19
	6	6	0.20	44	0.70
	20	50	0.96	215	0.96
	37	85	0.80	277	0.70
	50	64	0.48	192	0.36
Total lutein	-32	2	0.08	10	0.07
	6	4	0.15	29	0.21
	20	23	0.88	191	0.95
	37	85	0.95	432	0.95
	50	80	0.73	546	0.75
Total carotenoids	-32	0.8	0.01	10	0.07
	6	1	0.01	29	0.21
	20	33	0.88	192	0.95
	37	90	0.94	436	0.95
	50	83	0.71	550	0.75

Table 4. Reaction rate constant (k; month^{-1}), half-life ($t_{1/2}$; months), and D ($t_{1/10}$; months) for the total carotenoid content in durum wheat (Don Pedro variety) and tritordeum (HT621 line) whole-grain flours during a long-term storage period (12 months) at -32, 6, 20, 37, and 50 °C following the first-order kinetic model ($Ln(C/C_0) = -kt$).

Pigment	T (°C)	Durum Wheat (Don Pedro Variety)				Tritordeum (HT621 Advanced Line)			
		k ($\times 10^{-3}$ Month^{-1})	R^2	$t_{1/2}$ (Months)	$t_{1/10}$ (D) (Months)	k ($\times 10^{-3}$ Month^{-1})	R^2	$t_{1/2}$ (Months)	$t_{1/10}$ (D) (Months)
(all-*E*)-Zeaxanthin	-32	3	0.33	277	921	-	-	-	-
	6	5	0.79	141	470	-	-	-	-
	20	85	0.96	8	27	-	-	-	-
	37	252	0.97	3	9	-	-	-	-
	50	473	0.87	1	5	-	-	-	-
(all-*E*)-Lutein	-32	1	0.06	693	2302	2	0.15	385	1279
	6	3	0.18	217	719	12	0.72	60	198
	20	53	0.96	13	43	82	0.99	8	28
	37	191	0.97	4	12	329	0.99	2	7
	50	300	0.84	2	8	596	0.88	1	4
Total free (Z)-lutein	-32	2	0.04	385	1279	4	0.13	165	548
	6	0.7	0.00	990	3289	6	0.31	110	365
	20	38	0.93	18	60	39	0.98	18	59
	37	177	0.98	4	13	283	0.99	2	8
	50	249	0.88	3	9	559	0.99	1	4
(all-*E*)-β-Carotene	-32	5	0.35	139	460	1	0.02	576	1918
	6	2	0.06	462	1535	3	0.12	210	698
	20	45	0.90	16	52	38	0.95	19	62
	37	203	0.96	3	11	141	0.98	5	16
	50	387	0.95	2	6	417	0.99	2	6
Total free lutein	-32	1	0.07	693	2302	2	0.19	330	1096
	6	3	0.15	248	822	11	0.71	63	209
	20	51	0.97	14	45	76	0.99	9	30
	37	189	0.98	4	12	321	0.99	2	7
	50	290	0.85	2	8	630	0.94	1	4
Total free carotenoids	-32	0.7	0.01	990	3289	2	0.19	347	1151
	6	5	0.20	151	500	11	0.70	64	211
	20	53	0.96	13	44	75	0.99	9	31
	37	191	0.97	4	12	311	0.99	2	7
	50	296	0.85	2	8	636	0.94	1	4
Total lutein	-32	1	0.08	533	1771	1	0.07	495	1644
	6	3	0.15	248	822	4	0.21	178	590
	20	27	0.86	25	84	30	0.96	23	77
	37	145	0.98	5	16	103	0.98	7	22
	50	285	0.93	2	8	465	0.98	1	5
Total carotenoids	-32	0.5	0.00	1386	4604	1	0.07	495	1644
	6	0.9	0.01	770	2558	4	0.21	178	590
	20	30	0.87	23	78	30	0.96	23	77
	37	148	0.98	5	16	103	0.98	7	22
	50	290	0.93	2	8	466	0.98	1	5

The k values obtained for durum wheat were similar for xanthophylls and carotenes (β-carotene) with some exceptions (Table 4). (all-*E*)-Zeaxanthin was the pigment that was degraded most rapidly at all temperatures. As expected, the k value for zeaxanthin was maximum at 50 °C (473 × 10^{-3} month^{-1}), which is consistent with its complete disappearance at the seventh month in these conditions (Figure 2). Markedly, the differences between the k values for carotenes and xanthophylls were more evident in tritordeum flour. These results are in line with those obtained by Dhuique-Meyer et al. [45], whose study about the thermal degradation kinetics of vitamin C and carotenoids in citrus juices reported lower degradation rates for β-carotene than for xanthophylls.

The kinetic data obtained for the total carotenoids (which included the esterified pigments) confirmed a faster degradation in tritordeum than in durum wheat, especially at the higher storage temperatures, as indicated by the k values at 50 °C (Table 4). In the case of durum wheat flours, the k values at 20, 37, and 50 °C were consistently lower for total lutein and total carotenoids (including the xanthophyll esters) than for the respective free fractions, underlining the contribution of the esters to the greater stability of such fractions. Accordingly, this effect was more pronounced in tritordeum due to the higher content and proportion of esterified lutein.

The rate constants for lutein esters, including the distinction between regioisomers, are summarized in Table 5. In line with the results for total carotenoids (Table 4), the esterified fractions showed a higher degradation rate in tritordeum with compared to durum wheat. Notably, the k value for lutein diesters at 50 °C in tritordeum was double the k value in durum wheat. This result highlights an important turnover of diesters in durum wheat in accordance with a possible esterifying activity in this cereal. The differential content of other pro-oxidant and antioxidant substances in both cereal genotypes should also be considered. In any case, the k values decreased with an increase in the degree of esterification, with k lutein diester < k lutein monoester < k free lutein, resulting in an increased thermostability thereof. Regarding the regioisomers of the lutein monoesters (lutein monolinoleate and lutein monopalmitate), both positions 3 and 3' had similar degradation rates for each lutein monoester. No relevant differences between free and esterified lutein (Tables 4 and 5) were found in durum wheat, with the exception of the diesters. In tritordeum, lower degradation rates were recorded for lutein monoesters and diesters, even at 50 °C, despite the intense degradative conditions at that temperature. Lutein-3'-O-monolinoleate showed a degradation rate approximately 3-fold lower compared to free lutein at 37 °C. Within the monoester fraction, lutein monolinoleate presented a slower degradation than lutein monopalmitate; the same trend was observed for the monoesters at position 3' compared with the counterpart regioisomer at position 3. These data are consistent with the evolution described for the esterified fractions. In the case of the diesters, the degradation rates were even lower with no differences between the different acylated forms.

Table 5. Reaction rate constants (k; month^{-1}) for the esterified carotenoid content in durum wheat (Don Pedro variety) and tritordeum (HT621 line) whole-grain flours during a long-term storage period (12 months) at 20, 37, and 50 °C following the first-order kinetic model (Ln(C/C_0) = $-kt$).

Pigment	T (°C)	Durum Wheat (Don Pedro Variety)		Tritordeum (HT621 Advanced Line)	
		k ($\times 10^{-3}$ Month^{-1})	R^2	k ($\times 10^{-3}$ Month^{-1})	R^2
Lutein monolinoleate	20	-	-	-	-
	37	210	0.96	113	0.87
	50	262	0.96	388	0.90
Lutein-3'-O-linoleate	20	-	-	-	-
	37	184	0.94	103	0.91
	50	230	0.90	487	0.98
Lutein-3-O-linoleate	20	-	-	22	0.92
	37	200	0.95	140	0.94
	50	284	0.91	526	0.96
Lutein monopalmitate	20	-	-	34	0.97
	37	204	0.95	158	0.98
	50	332	0.94	544	0.96
Lutein-3'-O-palmitate	20	-	-	-	-
	37	201	0.95	147	0.97
	50	329	0.87	606	0.98
Lutein-3-O-palmitate	20	-	-	59	0.98
	37	186	0.96	228	1
	50	336	0.96	632	0.96
Total monoesters	20	-	-	13	0.78
	37	208	0.95	114	0.91
	50	310	0.94	471	0.95
Lutein-3,3'-dilinoleate	20	-	-	-	-
	37	111	0.90	85	0.92
	50	188	0.97	370	0.97
Lutein-3'-O-linoleate-3-O-palmitate plus Lutein-3'-O-palmitate-3-O-linoleate	20	-	-	-	-
	37	139	0.95	103	0.96
	50	201	0.96	400	0.97

Table 5. *Cont.*

Pigment	T (°C)	Durum Wheat (Don Pedro Variety)		Tritordeum (HT621 Advanced Line)	
		k ($\times 10^{-3}$ Month^{-1})	R^2	k ($\times 10^{-3}$ Month^{-1})	R^2
Lutein-3,3'-dipalmitate	20	-	-	-	-
	37	90	0.78	66	0.83
	50	156	0.96	451	0.98
Total diesters	20	-	-	-	-
	37	122	0.93	88	0.93
	50	190	0.98	395	0.97
Total esters	20	-	-	-	-
	37	162	0.96	-	-
	50	333	0.81	472	0.98

Half-life values ($t_{1/2}$; months) and D values ($t_{1/10}$; months) (Table 4) are a very useful tool for estimating the pigment concentration that will be retained in flours stored under controlled temperature conditions. Both parameters are inversely related to k values, so that an increase in the storage temperature results in a reduction in the $t_{1/2}$ and $t_{1/10}$ values. The half-life and D values were generally higher for durum wheat for all pigments at all storage temperatures. Thus, total free lutein and free carotenoids showed longer half-life values by 2 and 1 extra month, and D values of 5 and 4 extra months at 37 °C and 50 °C, respectively, for durum wheat compared to tritordeum. As an exception, the observed half-life and D values at 37 °C for total lutein and total carotenoids revealed the opposite situation, with tritordeum flour having higher values. These results are directly related to the higher content and proportion of esterified pigments in tritordeum, and they are in line with the lower carotenoid losses in tritordeum compared to durum wheat at 37 °C (Figure 1).

4. Conclusions

This comparative study evaluated the effects of processing and storage of whole-grain durum wheat and tritordeum flours on the total carotenoid content. The influence of the cereals' different genetic backgrounds was found to be important and the effect of storage was more severe on the carotenoid content of tritordeum. Tritordeum flour showed a lower retention of free and esterified carotenoids than durum wheat flour. This could be mediated by an increased esterifying activity in durum wheat flours and/or greater oxidative enzymatic activity, or a more oxidative environment in tritordeum flour. These results could be influenced by the fact that durum wheat varieties have generally suffered domestication and selective pressure by man for the preservation of the yellow colour trait. Our results suggest the occurrence of an enzymatic process, maybe a lipase, involved in the esterification of xanthophylls during the storage of these flours. The enzyme showed a preferential action for esterification of the hydroxyl group at position 3' in the ε-ring of the lutein molecule with linoleic acid. We hypothesize that this process could be different to the one described in intact grains in which the responsible enzymes (XAT: xanthophyll acyltransferase) showed a preferential acylation for the β-ring and a higher selectivity for palmitic acid, and therefore further research needs to be carried out to contrast this hypothesis. In any case, the increase in esterified xanthophylls eventually derived in a higher stability and retention capacity for total carotenoids in both cereal flours. This study provides valuable information to inform the optimization of storage conditions for flours of durum wheat and the novel hybrid cereal tritordeum with the aim of preserving their phytochemicals. This information could also be used in crop biofortification programs for the selection of cereal varieties, such as tritordeum, with an enhanced content of esterified xanthophylls.

Supplementary Materials: The following are available online at http://www.mdpi.com/2304-8158/6/12/111/s1, Figure S1: HPLC chromatograms obtained during long-term storage (12 months) at five different temperatures (−32, 6, 20, 37 and 50 °C) of durum wheat (Don Pedro) and tritordeum (HT621) whole-grain flours, Table S1: Evolution of the carotenoid composition (µg/g dry weight) during long-term storage (12 months) of durum wheat (Don Pedro variety) whole-grain flours, Table S2: Evolution of the carotenoid composition (µg/g dry weight) during long-term storage (12 months) of tritordeum (HT621 advanced line) whole-grain flours.

Acknowledgments: This work was supported by the Ministerio de Ciencia e Innovación (Spanish Government, Project AGL2010-14850/ALI) and the Consejería de Economía, Innovación, Ciencia y Empleo (Junta de Andalucía, Project P08-AGR-03477). E.M.-O. was the recipient of a JAE-Predoctoral grant (CSIC) co-financed by the ESF. D.H.-M. is member of CaRed Network funded by MINECO (BIO2015-71703-REDT) and EUROCAROTEN COST Action (CA15136). We acknowledge support of the publication fee by the CSIC Open Access Publication Support Initiative through its Unit of Information Resources for Research (URICI). We are grateful to Sergio G. Atienza (IAS-CSIC) for providing the plant material, and to Ruth Stuckey for language manuscript editing.

Author Contributions: D.H.-M. conceived and designed the experiments; E.M.-O. performed the experiments; D.H.-M. and E.M.-O. analyzed the data; D.H.-M. contributed reagents/materials/analysis tools; D.H.-M. and E.M.-O. wrote the paper.

Conflicts of Interest: The authors declare no conflict of interest.

References

1. Britton, G.; Liaaen-Jensen, S.; Pfander, H. *Carotenoids. Volume 1A: Isolation and Analysis*; Birkhäuser Verlag: Basel, Switzerland, 1995.

2. Howitt, C.A.; Pogson, B.J. Carotenoid accumulation and function in seeds and non-green tissues. *Plant Cell Environ.* **2006**, *29*, 435–445. [CrossRef] [PubMed]

3. Mayne, S.T. β-Carotene, carotenoids and disease prevention in humans. *FASEB J.* **1996**, *10*, 690–701. [PubMed]

4. Cooper, D.A. Carotenoids in health and disease: Recent scientific evaluations, research recommendations and the consumer. *J. Nutr.* **2004**, *134*, 221S–224S. [PubMed]

5. Britton, G.; Liaaen-Jensen, S.; Pfander, H. *Carotenoids Volume 5: Nutrition and Health*; Birkhäuser Verlag: Basel, Switzerland, 2009.

6. Landrum, J.T.; Bone, R.A. Lutein, zeaxanthin, and the macular pigment. *Arch. Biochem. Biophys.* **2001**, *385*, 28–40. [CrossRef] [PubMed]

7. Hentschel, V.; Kranl, K.; Hollmann, J.; Lindhauer, M.G.; Böhm, V.; Bitsch, R. Spectrophotometric determination of yellow pigment content and evaluation of carotenoids by high-performance liquid chromatography in durum wheat grain. *J. Agric. Food Chem.* **2002**, *50*, 6663–6668. [CrossRef] [PubMed]

8. Ficco, D.B.M.; Mastrangelo, A.M.; Trono, D.; Borrelli, G.M.; De Vita, P.; Fares, C.; Beleggia, R.; Platani, C.; Papa, R. The colours of durum wheat: A review. *Crop Pasture Sci.* **2014**, *65*, 1–15. [CrossRef]

9. Bai, C.; Twyman, R.M.; Farré, G.; Sanahuja, G.; Christou, P.; Capell, T.; Zhu, C.A. Golden era—Provitamin A enhancement in diverse crops. *In Vitro Cell. Dev. Biol. Plant* **2011**, *47*, 205–221. [CrossRef]

10. Martín, A.; Sanchez-Monge, E.L. Citology and morphology of the amphiploid *Hordeum chilense-Triticum turgidum* conv. *Durum. Euphytica* **1982**, *31*, 261–267. [CrossRef]

11. Atienza, S.G.; Ballesteros, J.; Martín, A.; Hornero-Méndez, D. Genetic variability of carotenoid concentration and degree of esterification among tritordeum (×*Tritordeum* Ascherson et Graebner) and durum wheat accessions. *J. Agric. Food Chem.* **2007**, *55*, 4244–4251. [CrossRef] [PubMed]

12. Mellado-Ortega, E.; Hornero-Méndez, D. Isolation and identification of lutein esters, including their regioisomers, in tritordeum (×*Tritordeum* Ascherson et Graebner) grains. Evidences for a preferential xanthophyll acyltransferase activity. *Food Chem.* **2012**, *135*, 1344–1352. [CrossRef] [PubMed]

13. Mellado-Ortega, E.; Hornero-Méndez, D. Carotenoid profiling of *Hordeum chilense* grains: The parental proof for the origin of the high carotenoid content and esterification pattern of tritordeum. *J. Cereal Sci.* **2015**, *62*, 15–21. [CrossRef]

14. Chavez, A.L.; Sanchez, T.; Ceballos, H.; Rodriguez-Amaya, D.B.; Nestel, P.; Tohme, J.; Ishitani, M. Retention of carotenoids in cassava roots submitted to different processing methods. *J. Sci. Food Agric.* **2007**, *87*, 388–393. [CrossRef]

15. Mugode, L.; Ha, B.; Kaunda, A.; Sikombe, T.; Phiri, S.; Mutale, R.; Davis, C.; Tanumihardjo, S.; De Moura, F. Carotenoid retention of biofortified provitamin A maize (*Zea mays* L.) after Zambian traditional methods of milling, cooking and storage. *J. Agric. Food Chem.* **2014**, *62*, 6317–6325. [CrossRef] [PubMed]

16. Fardet, A. New hypotheses for the health-protective mechanisms of wholegrain cereals: What is beyond fibre? *Nutr. Res. Rev.* **2010**, *23*, 65–134. [CrossRef] [PubMed]

17. Britton, G.; Khachik, F. Carotenoids in food. In *Carotenoids Volume 5: Nutrition and Health*; Britton, G., Liaaen-Jensen, S., Pfander, H., Eds.; Birkhäuser Verlag: Basel, Switzerland, 2009; pp. 45–66.

18. De Moura, F.F.; Miloff, A.; Boy, E. Retention of provitamin A carotenoids in staple crops targeted or biofortification in Africa: Cassava, maize, and sweet potato. *Crit. Rev. Food Sci. Nutr.* **2015**, *55*, 1246–1269. [CrossRef] [PubMed]

19. Hemery, Y.; Rouau, X.; Lullien-Pellerin, V.; Barron, C.; Abecassis, J. Dry processes to develop wheat fractions and products with enhanced nutritional quality. *J. Cereal Sci.* **2007**, *46*, 327–347. [CrossRef]

20. Li, L.; Yong, Y.; Qiang, X.; Owsiany, K.; Welsch, R.; Chitchumroonchokchai, C.; Lu, S.; Van Eck, J.; Deng, X.; Failla, M.; et al. The *Or* gene enhances carotenoid accumulation and stability during post-harvest storage of potato tubers. *Mol. Plant* **2012**, *5*, 339–352. [CrossRef] [PubMed]

21. Ortiz, D.; Rocheford, T.; Ferruzzi, M.G. Influence of temperature and humidity on the stability of carotenoids in biofortified maize (*Zea mays* L.) genotypes during controlled postharvest storage. *J. Agric. Food Chem.* **2016**, *64*, 2727–2736. [CrossRef] [PubMed]

22. Ahmad, F.T.; Asenstorfer, R.E.; Soriano, I.R.; Mares, D.J. Effect of temperature on lutein esterification and lutein stability in wheat grain. *J. Cereal Sci.* **2013**, *58*, 408–413. [CrossRef]

23. Mellado-Ortega, E.; Hornero-Méndez, D. Carotenoid evolution during short-storage period of durum wheat (*Triticum turgidum* conv. *durum*) and tritordeum ($\times$ *Tritordeum* Ascherson et Graebner) whole-grain flours. *Food Chem.* **2016**, *192*, 714–723. [CrossRef] [PubMed]

24. Mellado-Ortega, E.; Hornero-Méndez, D. Effect of long-term storage on free and esterified carotenoids in durum wheat (*Triticum turgidum* conv. *durum*) and tritordeum ($\times$ *Tritordeum* Ascherson et Graebner) grains. *Food Res. Int.* **2017**, *99*, 877–890. [CrossRef] [PubMed]

25. Ballesteros, J.B.; Ramírez, M.C.; Martínez, C.; Atienza, S.G.; Martín, A. Registration of HT621, a high carotenoid content tritordeum germplasm line. *Crop Sci.* **2005**, *45*, 2662–2663. [CrossRef]

26. Mellado-Ortega, E.; Atienza, S.G.; Hornero-Méndez, D. Carotenoid evolution during postharvest storage of durum wheat (*Triticum turgidum* conv. *durum*) and tritordeum ($\times$ *Tritordeum* Ascherson et Graebner) grains. *J. Cereal Sci.* **2015**, *62*, 134–142. [CrossRef]

27. Mínguez-Mosquera, M.I.; Hornero-Méndez, D. Separation and quantification of the carotenoid pigments in red peppers (*Capsicum annuum* L.), paprika and oleoresin by reversed-phase HPLC. *J. Agric. Food Chem.* **1993**, *43*, 1613–1620. [CrossRef]

28. Britton, G. UV/visible spectroscopy. In *Carotenoids. Volume 1B: Spectroscopy*; Britton, G., Liaaen-Jensen, S., Pfander, H., Eds.; Birkhäuser Verlag: Basel, Switzerland, 1995; pp. 13–62.

29. Upadhyay, S.K. Elementary. In *Chemical Kinetics and Reaction Dynamics*; Anamaya Publishers: New Delhi, India, 1996; pp. 1–45.

30. Gayen, D.; Ali, N.; Sarkar, S.N.; Datta, S.K.; Datta, K. Down-regulation of lipoxygenase gene reduces degradation of carotenoids of golden rice during storage. *Planta* **2015**, *242*, 353–363. [CrossRef] [PubMed]

31. Rani, K.U.; Prasada Rao, U.J.S.; Leelavathi, K.; Haridas Rao, P. Distribution of enzymes in wheat flour mill streams. *J. Cereal Sci.* **2001**, *34*, 233–242. [CrossRef]

32. Li, D.; Song, J.; Liu, C. Kinetic stability of lutein in freeze-dried sweet corn powder stored under different conditions. *Food Sci. Technol. Res.* **2014**, *20*, 65–70. [CrossRef]

33. Choe, E.; Lee, J.; Park, K.; Lee, S. Effects of heat pretreatment on lipid and pigments of freeze-dried spinach. *J. Food Sci.* **2001**, *66*, 1074–1079. [CrossRef]

34. Urquhart, A.A.; Altosaar, I.; Matlashewski, G.J.; Sahasrabudhe, M.R. Localization of lipase activity in oat grains and milled oat fractions. *Cereal Chem.* **1983**, *60*, 181–183.

35. Barros, M.; Fleuri, L.F.; Macedo, G.A. Seed lipases: Sources, applications and properties—A review. *Braz. J. Chem. Eng.* **2010**, *27*, 15–29. [CrossRef]

36. Farrington, F.F.; Warwick, M.J.; Shearer, G. Changes in the carotenoids and sterol fractions during the prolonged storage of wheat flour. *J. Sci. Food Agric.* **1981**, *32*, 948–950. [CrossRef]

37. Ahmad, F.T.; Mather, D.E.; Law, H.; Li, M.; Yousif, S.; Chalmers, K.J.; Asenstorfer, R.E.; Mares, D.J. Genetic control of lutein esterification in wheat (*Triticum aestivum* L.) grain. *J. Cereal Sci.* **2015**, *64*, 109–115. [CrossRef]

38. Mattera, G.; Cabrera, A.; Hornero-Méndez, D.; Atienza, S.G. Lutein esterification in wheat endosperm is controlled by the homoeologous group 7, and is increased by the simultaneous presence of chromosomes 7D and 7H[ch] from *Hordeum chilense*. *Crop Pasture Sci.* **2015**, *66*, 912–921. [CrossRef]

39. Mattera, G.; Hornero-Méndez, D.; Atienza, S.G. Lutein ester profile in wheat and tritordeum can be modulated by temperature: Evidences for regioselectivity and fatty acid preferential of enzymes encoded by genes on chromosomes 7D and 7H[ch]. *Food Chem.* **2017**, *219*, 199–206. [CrossRef] [PubMed]

40. O'Connor, J.; Perry, H.J.; Harwood, J.L. A comparison of lipase activity in various cereal grains. *J. Cereal Sci.* **1992**, *16*, 153–163. [CrossRef]

41. Hidalgo, A.; Brandolini, A. Kinetics of carotenoids degradation during the storage of einkorn (*Triticum monococcum* L. ssp. *monococcum*) and bread wheat (*Triticum aestivum* L. ssp. *aestivum*) flours. *J. Agric. Food Chem.* **2008**, *56*, 11300–11305. [CrossRef] [PubMed]

42. Borrelli, G.M.; Troccoli, A.; Di Fonzo, N.; Fares, C. Durum wheat lipoxygenase activity and other quality parameters that affect pasta colour. *Cereal Chem.* **1999**, *76*, 335–340. [CrossRef]

43. Leenhardt, F.; Lyan, B.; Rock, E.; Boussard, A.; Potus, J.; Chanliaud, E.; Remesy, C. Wheat lipoxygenase activity induces greater loss of carotenoids than vitamin E during breadmaking. *J. Agric. Food Chem.* **2006**, *54*, 1710–1715. [CrossRef] [PubMed]

44. Fratianni, A.; Di Criscio, T.; Mignogna, R.; Panfili, G. Carotenoids, tocols and retinols evolution during egg pasta-making processes. *Food Chem.* **2012**, *131*, 590–595. [CrossRef]

45. Dhuique-Mayer, C.; Tbatou, M.; Carail, M.; Caris-Veyrat, C.; Dornier, M.; Amiot, M.J. Thermal degradation of antioxidant micronutrients in citrus juice: Kinetics and newly formed compounds. *J. Agric. Food Chem.* **2007**, *55*, 4209–4216. [CrossRef] [PubMed]

foods

MDPI

Article

Production of Barbari Bread (Traditional Iranian Bread) Using Different Levels of Distillers Dried Grains with Solubles (DDGS) and Sodium Stearoyl Lactate (SSL)

Shirin Pourafshar [1], Kurt A. Rosentrater [2,3,*] and Padmanaban G. Krishnan [1]

[1] Dairy and Food Science Department, South Dakota State University, Brookings, SD 57007, USA;
    SP8DS@hscmail.mcc.virginia.edu (S.P.); Padmanaban.Krishnan@sdstate.edu (P.G.K.)
[2] Department of Agriculture and Biosystems Engineering, Iowa State University, Ames, IA 50011, USA
[3] Department of Food Science and Human Nutrition, Iowa State University, Ames, IA 50011, USA
* Correspondence: karosent@iastate.edu; Tel.: +1-515-294-4019

Received: 22 November 2017; Accepted: 27 February 2018; Published: 1 March 2018

Abstract: Bread is one of the oldest foods known throughout history and even though it is one of the principal types of staple around the world, it usually lacks enough nutrients, including protein and fiber. As such, fortification is one of the best solutions to overcome this problem. Thus, the objective this study was to examine the effect of three levels of distillers dried grains with solubles (DDGS) (0%, 10% and 20%) in conjunction with three levels of SSL (sodium stearoyl lactate) (0%, 2% and 5%) on physical and chemical properties of Barbari bread (traditional Iranian bread). To the best of our knowledge, this is the first study to evaluate DDGS and Sodium Stearoyl-2-Lactate (SSL), as sources of fortification in Barbari bread. The results showed that incorporation of 20% of DDGS and 0% SSL caused a significant increase in the amount of fiber and protein. As for the physical attributes, using higher amount of DDGS caused a darker color, and as for the texture parameters, the highest firmness was measured when 10% DDGS and 5% of SSL were used. Different Mixolab and Rapid Visco Analyzer (RVA) parameters also were measured with varying results. The findings of this study show that DDGS can be a valuable source of fiber and protein, which can be used as a cost effective source to fortify cereal-based products.

Keywords: distillers dried grains with solubles; fortification; sodium stearoyl sactate

1. Introduction

Cereals are the edible seeds or grains of the grass family, *Gramineae.* Because cereals are inexpensive and readily available, humans in almost every country have used them as major food staples for centuries. Cereals and cereal products are an important source of energy, protein and fiber [1]. Wheat is the most important cereal, and is commonly consumed worldwide. Historians do not know exactly where wheat was first cultivated, but sources point to either Syria-Palestine or southern parts of Anatolia. Wheat cultivation spread from Palestine to Egypt and then from northern Mesopotamia to Persia, where bread was first developed. From there, the growth of wheat and bread spread in all directions [2]. Although whole wheat as a food component has high nutritional value, a considerable proportion of the grain's nutrients are lost during the milling processes. Thus, the importance of adding value to those products made with all-purpose flour or other wheat flours increases. Since bread is the most consumed cereal product, fortification can help combat problems such as malnutrition.

One way of fortifying cereal products, especially breads, may be through the use of distillers dried grains. Distillers dried grains with solubles (DDGS) is a co-product resulting from the fermentation of cereal grains, mostly corn, for the production of ethanol. As a result of increase in ethanol production, there is an increase supply of DDGS as well [3]. America's corn farmers have improved their cultivation techniques significantly and increased their yield potential. In 1935, 82 million acres of corn were harvested in United States (U.S.) with an average yield of 24.2 bushels per acre; by 1950, the yield increased to 38.2 bushels per acre. Then in 1956, the problem with farms was the abundance of corn. Corn production continued increasing so that the yield was 149 bushels per acre by 2006. The increase in corn encouraged cattle feeding in the U.S, and with the growth of the ethanol industry, the demand for corn has improved [4]. Furthermore, researchers have identified the potential value of DDGS as a source of protein, which often ranges from 27% to 35%, fiber, minerals and vitamins [5]. As a result, scientists and engineers have been trying to find different ways of using DDGS in human foods, rather than solely as livestock feed. Researchers have explored incorporating DDGS in food products, especially cereal-based products. For instance, in a study done by Wu et al. [6], spaghetti was supplemented with corn distillers dried grains. Additionally, Finley and Hanamoto [7] used brewer's spent grain in bread, while Tsen et al. [8,9] used DDG flour in the production of bread and cookies and then evaluated the physical and chemical properties of final products. Corn distiller's grains have also been used in spaghetti [8]. Furthermore, the effect of DDGS on quality of cornbread has been investigated by Liu and colleagues [10].

Sodium Stearoyl-2-Lactilate (SSL) is an anionic emulsifier which is effective in increasing dough strength. Emulsifiers are mostly used in the baking industry to enhance baking quality. They can prevent mechanical damage to fermented dough, increase shelf life and improve the texture of baked products [11]. Hydrocolloids are often used in bread to improve the volume and texture, extend shelf life, and make softer bread crumbs [12]. SSL can function through interactions with flour protein which will improve the viscoelasticity of the dough [11]. Other protein-reactive softeners, such as DATEM (diacetyl tartaric ester of monoglycerides), can increase the strength of gluten protein matrix which will, in turn, improve loaf volume and tighten crumb structure [13].

Flat breads have a very short shelf life, usually a few hours. Because of that, many studies have been done to increase the shelf life of flat breads. For example, in a study by Qarooni [2], the anti-staling effect of ingredients such as shortening on Barbari bread's quality was investigated. The results from that study showed that adding 0.5% SSL and 0.3% shortening made the bread edible for up to 36 h, instead of the normal 16 h. Different Middle Eastern breads are made with various types of flours. For example, round shaped Baladi and Aish Meharha from Egypt and Bazlama, Pide and Yufca from Turkey are mainly made with wheat flour. Morocco has pan fried bread made with semolina flour. Afghanistan and Tajikistan have Bolani bread, which is flat bread stuffed with different vegetables. Although these breads are mainly made with wheat and other types of flours, each of them has its own physical and chemical characteristics. However, they may have deficiencies in certain nutrient components that can be remedied through fortification.

Among Middle Easterners, Iranians consume four major types of breads: Barbari, Lavash, Sangak and Taftoon, with Barbari being the most popular. On average, Barbari crust has a thickness of 1–2 mm, length of 67–75 cm, and width of 13.5–20 cm. Traditionally, this bread is made with all-purpose flour and the final product contains about 11% protein, 10% fiber and 0.5% fat. Barbari is thick and oval shaped, and is often topped with poppy seeds. The golden color of this bread comes from Romal, a mixture made from flour, baking soda, and boiling water which is brushed on the dough before baking. Barbari has a special aroma and its taste depends on the amount of sour dough and baking time.

In this study, three levels of DDGS (0%, 10% and 20%) and three levels of SSL (0%, 2% and 5%) were used for substitution of wheat flour in Barbari bread. The objectives of this study were to understand (1) the impact of substitution of three levels of DDGS and SSL on the physical and chemical attributes of final bread products, and (2) to study the changes in the physical properties of the dough with different levels of substitutions of DDGS and SSL.

2. Materials and Methods

2.1. Experimental Design

Distillers dried grains with solubles (DDGS) was obtained from a commercial fuel ethanol plant in South Dakota. All-purpose wheat flour and other ingredients were purchased from local markets. A control sample of Barbari bread was baked at 550 °C for 10 min. To fortify this bread three levels of DDGS substitution (0%, 10% and 20%) and three levels of SSL substitution (0%, 2% and 5%) were used, producing a two-factor design each with three levels, having a full factorial design. Two loaves of breads for each level of substitution were baked. All properties of each of these loaves were analyzed using three replications; thus, $n = 6$ measurements for each property, and for each treatment combination. In total, 54 samples were analyzed. Table 1 shows the amount of all-purpose flour, DDGS and SSL to be used in samples.

Table 1. Experimental design [1].

Treatment	Wheat [2] (%)	DDGS [3] (%)	SSL [4] (%)
0 (Control)	100	0	0
1	98	0	2
2	95	0	5
3	90	10	0
4	88	10	2
5	85	10	5
6	80	20	0
7	78	20	2
8	75	20	5

[1] Each treatment was replicated twice. [2] From market source, Hy-Vee bleached all-purpose flour. [3] DDGS is distillers dried grains with solubles, obtained from a commercial fuel ethanol plant. [4] SSL is sodium stearoyl-2-lactate.

2.2. Preparation

For the sour dough, 1 g of salt, 9 g of active dry yeast (Red Star Active Dry Yeast, purchased from local market), 400 g of flour (Gold Medal All-purpose Flour, purchased from local market) with 650 g of water were used and only 35 g of sour dough was used for each batch. Romal topping for the bread required 4.2 g of flour, 4.2 g of baking soda and 85 g of water.

For the control, bread was made with only wheat flour, 880 g of all-purpose flour (Gold Medal All-purpose Flour) was used; the rest of the ingredients were 4.2 g of sugar, 4.2 g of salt, and 689.76 g of water. For the other breads, the same ingredients were used, varying only the amount of flour.

2.3. Bread Production

Sour dough was prepared 18 h before bread preparation during which, the sour dough was covered and left at room temperature. In terms of bread preparation, first the yeast was dissolved in warm water and then sugar was added and put aside for 10 min. This was then mixed with salt and water and then flour was gradually added and sour dough from the previous day was added as well. The mixture was mixed until the dough was no longer sticky. The next step was proofing, where the dough was placed in a proofing chamber for an hour and half for further activity of the yeast. Then, 400 g of dough was weighed and kneaded to form a 20 inch (50.8 cm) by 20 inch (50.8 cm) square, using a square frame to assure consistency in dimension of all breads. The thickness of the dough was measured in three different areas at the edges, then the Romal was made and brushed on top of dough. The dough was put aside for 10 min. Next, bread was baked at 500 °C for 10 min on a pizza stone (14 inch by 16 inch) to make baking condition close to that of the traditional ovens used for Barbari in Iran.

The other breads were made the same way except for the amount of flour, DDGS and SSL which were incorporated accordingly in each bread sample (Table 1). In the breads other than control, the same procedure was used but with different proportions of all-purpose flour. The mixing time and other details of preparation were the same for control and all other breads.

2.4. Physical and Chemical Properties

A texture analyzer was used to study the firmness and extensibility of the bread samples using two different probes: SMS/Chen-Hoseney Dough stickiness RIG and Pizza Tensile RIG, (TX.XT-plus, Texture Technologies Corp., Scarsdale, NY, USA). For measurement of each of these variables, duplications were done for each loaf ($n = 2$), for a total of four samples for each type of bread, or 24 samples. The thickness at the edges and the center of the bread was measured using Vernier calipers (Digimatic Calipers w/Absolute Encoders, Series 500, Mitutoyo Corporation, Kawasaki, Kanagawa, Japan). For the center, because of the bubbles which were formed in the middle of breads, the thickness was measured three times. After that, breads were grinded and moisture content as well as water activity were determined. In order to determine the water activity, water activity meter for food quality was used (Aqua lab CX-2, Decagon Devices, Inc., Pullman, WA, USA). Color was measured by spectrophotometer (Minolta CM-508d, Ramsey, NJ, USA) in which L^* is the measure of lightness, a^* is the measure of greenness to redness, and b^* is the measure of blueness to yellowness. The color was determined for the baked products just like it was done for the dough, and L^*, a^* and b^* were measured to get the color values.

Protein content was measured using the American Association of Cereal Chemistry (AACC) method for combustion—AACC approved method 46-0 [14] with a CE Elantech instrument (Flash EA 1112, ThermoFinnigan Italia S.p.A., Rodano (MI) Italy). In this method, the amount of nitrogen which was determined by the machine was converted into protein using a conversion factor of 5.7. For the determination of neutral detergent fiber (NDF), AACC approved method 30-25 (2010) was used. Fat content was determined using AOAC method 920.39 (2003) with an automated extractor Soxhlet using petroleum ether (CH-9230, Buchi Laborotechnik AG, Flawil, Switzerland). Moisture content was determined using the AACC approved method 44-19 [15] convection oven drying at 135°C (Model Labline, Inc. Chicago, IL, USA).

The rheological properties of the dough were determined using a Mixolab (Tripette and Renaud Chopin, Villeneuve La Garenne cedex, France) and Rapid Visco Analyzer (RVA) (Newport Scientific Pty. Ltd., Warriewood, Australia). For the Mixolab, the minimum torque (C_2), peak torque (C_3), cooking stability (C_4), set back (C_5) and the α, β, and γ were evaluated. As for the RVA, peak viscosity, temperature at peak viscosity, time to peak viscosity, and breakdown were measured.

2.5. Data Analysis

All collected data were analyzed with Microsoft Excel v.2007 and SAS v.9.0 software (SAS Institute, Cary, NC, USA) using a Type I error rate (α) of 0.05, by analysis of variance (ANOVA) to identify significant differences among treatments. Post-hoc Fisher's Least Significant Differences LSD tests were used to determine where the differences occurred.

3. Results and Discussion

The results for each measurement are summarized in following tables: Table 2 shows the effect of adding DDGS or SSL on chemical properties of final breads; Table 3 shows the effect of adding both DDGS and SSL on chemical properties of final breads; Table 4 shows the main effect of adding DDGS or SSL on physical properties of final breads; and Table 5 describes the effect of adding both DDGS and SSL on physical attributes of final breads. The results from treatment effects on Mixolab and RVA are shown in Tables 6–10, respectively. Tables 11 and 12 show the results for the main effect and treatment effect of DDGS and SSL on bread quality, respectively.

Table 2. Main effects on chemical properties of baked breads [1].

Effect (% Substitution)	Protein (% db) [2]		Fiber (% db)		Fat (% db)		Moisture	
	Mean	**SD** [3]	**Mean**	**SD**	**Mean**	**SD**	**Mean**	**SD**
DDGS (%)								
0	11.10 [a]	0.23	0.45 [c]	0.37	1.22 [c]	0.84	0.43 [ab]	0.03
10	12.48 [b]	0.15	2.05 [b]	0.37	1.75 [b]	1.09	0.40 [b]	0.35
20	13.66 [c]	0.10	3.71 [a]	0.42	2.4 [a]	1.12	0.38 [a]	0.57
SSL (%)								
0	12.51 [a]	1.17	2.42 [a]	1.44	0.86 [c]	0.40	0.44 [a]	0.04
2	12.49 [a]	1.04	2.04 [b]	1.59	1.49 [b]	0.59	0.42 [a]	0.06
5	12.24 [b]	1.21	1.75 [b]	1.38	3.07 [a]	0.68	0.42 [a]	0.04

[1] Different letters for a given dependent variable denote significant differences ($\alpha = 0.05$) across treatment conditions for that independent variable. [2] All properties are reported as % dry basis (db). [3] SD is standard deviation.

Table 3. Treatment effects on chemical properties of baked breads [1].

DDGS (%)	SSL (%)	Protein (% db) [2]		Fiber (% db)		Fat (% db)		Moisture (% db)	
		Mean	**SD** [3]	**Mean**	**SD**	**Mean**	**SD**	**Mean**	**SD**
0	0	11.17 [d]	0.04	0.90 [d]	0.04	0.50 [f]	0.00	0.45 [a]	0.02
0	2	11.29 [d]	0.24	0.28 [e]	0.25	0.87 [e]	0.03	0.43 [ab]	0.03
0	5	10.86 [e]	0.12	0.17 [e]	0.05	2.29 [c]	0.10	0.40 [ab]	0.01
10	0	12.58 [b]	0.05	2.26 [c]	0.36	0.73 [ef]	0.05	0.41 [ab]	0.00
10	2	12.58 [b]	0.01	2.02 [c]	0.05	1.43 [d]	0.01	0.38 [b]	0.01
10	5	12.29 [c]	0.07	1.88 [c]	0.64	3.11 [b]	0.06	0.41 [ab]	0.07
20	0	13.79 [a]	0.05	4.09 [a]	0.21	1.35 [d]	0.20	0.47 [a]	0.05
20	2	13.62 [a]	0.02	3.84 [a]	0.00	2.19 [c]	0.13	0.46 [a]	0.08
20	5	13.57 [a]	0.056	3.20 [b]	0.07	3.82 [a]	0.15	0.46 [a]	0.03

[1] Different letters for a given dependent variable denote significant differences ($\alpha = 0.05$) across treatment conditions for that independent variable. [2] All properties are reported as % dry basis. [3] SD is standard deviation.

Table 4. Main effects on physical properties of baked breads [1].

Effect (% Substitution)	Firmness (N)[2] Mean	SD[3]	Extensibility (N)[2] Mean	SD	Thickness (mm) Mean	SD	a_w Mean	SD	L^* Mean	SD	a^* Mean	SD	b^* Mean	SD
DDGS (%)														
0	31.28 b	8.98	19.06 a	8.08	13.97 b	3.24	0.93 a	0.01	61.6 b	5.88	9.78 a	3.32	20.02 a	4.67
10	42.62 a	8.13	13.15 b	6.14	14.80 ab	1.01	0.88 a	0.16	72.94 a	2.68	10.22 a	2.79	22.86 a	2.15
20	39.10 a	5.41	10.41 b	2.99	16.06 a	1.84	0.94 a	0.00	71.03 a	4.01	9.74 a	1.57	23.67 a	1.84
SSL (%)														
0	33.57 b	10.91	17.65 a	4.60	13.32 b	2.86	0.94 a	0.01	69.51 a	4.77	11.01 a	2.55	23.27 a	1.71
2	38.85 ab	5.12	15.55 a	9.32	15.67 a	1.46	0.89 a	0.17	67.95 a	9.39	8.63 a	2.35	20.60 a	4.85
5	40.57 a	8.68	9.42 b	2.56	15.85 a	1.64	0.93 a	0.013	68.13 a	5.78	10.11 a	2.43	22.68 a	2.73

[1] Different letters for a given dependent variable denote significant differences ($\alpha = 0.05$) across treatment conditions for that independent variable. [2] The force was measured over time for firmness, and over travel distance for extensibility. [3] SD is standard deviation. a_w is water activity; L^* is the measure of lightness; a^* is the measure of greenness to redness; and b^* is the measure of blueness to yellowness.

Table 5. Treatment effects on physical properties of baked breads [1].

DDGS (%)	SSL (%)	Firmness (N)[2] Mean	SD[3]	Extensibility (N)[2] Mean	SD	Thickness (mm) Mean	SD	a_w Mean	SD	L^* Mean	SD	a^* Mean	SD	b^* Mean	SD
0	0	20.95 d	5.83	18.53 b	5.28	9.70 d	0.97	0.94 a	0.00	65.14 cb	6.56	13.54 a	2.02	25.13 a	1.27
0	2	38.75 bc	5.13	27.05 a	7.03	15.72 abc	0.69	0.94 a	0.00	55.97 d	2.34	6.70 b	1.71	14.99 c	2.19
0	5	34.15 c	2.83	11.59 cd	1.49	16.49 ab	0.56	0.92 b	0.00	63.74 cd	5.21	9.11 ab	0.28	19.95 b	0.11
10	0	42.36 bc	4.50	21.05 b	2.71	14.72 bc	0.98	0.93 a	0.00	70.97 abc	3.33	11.24 ab	0.26	23.02 ab	0.85
10	2	35.19 bc	3.89	9.18 cd	1.51	0.38 c	0.01	0.92 b	1.30	74.62 a	1.44	8.98 ab	3.18	22.21 ab	2.98
10	5	50.33 a	7.34	9.23 cd	1.97	15.24 abc	0.79	0.93 a	0.02	73.23 ab	3.04	10.45 ab	4.85	23.35 ab	3.50
20	0	37.42 bc	6.85	13.38 c	1.25	15.55 abc	1.22	0.95 a	0.00	72.41 abc	0.75	8.25 b	0.67	21.68 ab	0.57
20	2	42.63 b	4.20	10.43 cd	1.23	16.83 a	1.36	0.93 a	0.00	73.28 ab	1.01	10.21 ab	1.41	24.61 a	1.95
20	5	37.24 bc	4.21	7.44 d	2.51	15.80 abc	2.81	0.94 a	0.00	67.41 abc	6.24	10.76 ab	1.73	24.73 a	0.94

[1] Different letters for a given dependent variable denote significant differences ($\alpha = 0.05$) across treatment conditions for that independent variable. [2] The force was measured over time for firmness, and over travel distance for extensibility. [3] SD is standard deviation. a_w is water activity; L^* is the measure of lightness; a^* is the measure of greenness to redness; and b^* is the measure of blueness to yellowness.

Table 6. Main effect of DDGS on Mixolab operational parameters [1].

Effect (% Substitution)	C1 Time (min)		C1 Torque (Nm)		C2 Torque (Nm)		C3 Torque (Nm)		C4 Torque (Nm)		C5 Torque (Nm)		α^4 (N-m/min)		β^4 (N-m/min)		γ^4 (N-m/min)		Water Abs [3] (%)		Stability (s)	
	Mean	SD [2]	Mean	SD	Mean	SD	Mean	SD	Mean	SD	Mean	SD	Mean	SD	Mean	SD	Mean	SD	Mean	SD	Mean	SD
0	2.15 a	2.23	1.09 b	0.03	0.37 a	0.06	1.49 a	0.52	1.66 a	0.35	2.34 a	0.49	−0.05 a	0.03	0.31 ab	0.26	−0.01 a	0.08	53.36 a	2.88	7.60 a	4.12
10	2.47 a	2.29	1.12 ab	0.03	0.36 a	0.05	1.16 a	0.46	1.29 a	0.75	2.23 a	0.95	−0.06 a	0.03	0.37 a	0.31	−0.01 a	0.06	53.13 a	2.05	5.77 a	3.90
20	3.90 a	3.48	1.16 a	0.07	0.39 a	0.06	0.69 b	0.15	1.01 a	0.67	2.20 a	0.94	−0.07 a	0.03	0.093 b	0.034	0.02 a	0.02	52.08 a	2.12	5.80 a	3.70

[1] Different letters for a given dependent variable denote significant differences (α = 0.05) across treatment conditions for that independent variable. [2] SD is standard deviation. [3] Abs is absorption. [4] α shows protein breakdown, β shows gelatinization and γ shows cooking stability rate.

Table 7. Main effect of SSL on Mixolab operational parameters [1].

Effect (% Substitution)	C1 Time (Min)		C1 Torque (Nm)		C2 Torque (Nm)		C3 Torque (Nm)		C4 Torque (Nm)		C5 Torque (Nm)		α^4 (N-m/min)		β^4 (N-m/min)		γ^4 (N-m/min)		Water Abs [3] (%)		Stability (s)	
	Mean	SD [2]	Mean	SD	Mean	SD	Mean	SD	Mean	SD	Mean	SD	Mean	SD	Mean	SD	Mean	SD	Mean	SD	Mean	SD
0	4.33 a	2.34	1.13 a	0.08	0.35 b	0.03	1.31 a	0.54	1.52 a	0.11	2.27 a	0.30	−0.08 b	0.04	0.33 a	0.20	0.00 a	0.04	53.13 a	2.41	10.22 a	1.11
2	3.17 ab	3.48	1.11 a	0.05	0.42 ab	0.05	1.20 ab	0.52	1.51 a	0.54	2.61 a	0.43	−0.06 ab	0.02	0.37 a	0.32	0.00 a	0.06	53.45 a	2.71	5.74 b	3.49
5	1.02 b	0.14	1.13 a	0.03	0.35 a	0.06	0.83 b	0.41	0.93 a	0.90	1.89 a	1.20	−0.04 a	0.02	0.07 b	0.07	0.00 a	0.08	52.00 a	1.89	3.20 b	2.20

[1] Different letters for a given dependent variable denote significant differences (α = 0.05) across treatment conditions for that independent variable. [2] SD is standard deviation. [3] Abs is absorption. [4] α shows protein breakdown, β shows gelatinization and γ shows cooking stability rate.

Table 8. Treatment effects on Mixolab operational parameters [1].

DDGS (%)	SSL (%)	C1 Time (Min)		C1 Torque (Nm)		C2 Torque (Nm)		C3 Torque (Nm)		C4 Torque (Nm)		C5 Torque (Nm)		α^3 (N-m/min)		β^3 (N-m/min)		γ^3 (N-m/min)		Water Abs [3] (%)		Stability (s)	
		Mean	SD [2]	Mean	SD	Mean	SD	Mean	SD	Mean	SD	Mean	SD	Mean	SD	Mean	SD	Mean	SD	Mean	SD	Mean	SD
0	0	1.43 ab	0.12	1.08 a	0.00	0.37 ab	0.04	1.73 a	0.14	1.44 bc	0.09	2.00 ab	0.13	−0.04 abc	0.04	0.50 ab	0.01	−0.05 bc	0.01	50.75 b	2.05	10.55 a	1.45
0	2	3.99 ab	3.83	1.06 a	0.01	0.37 ab	0.05	1.71 a	0.13	1.44 bc	0.16	2.13 ab	0.36	−0.08 abc	0.03	0.47 abc	0.04	−0.05 ab	0.07	56.05 a	1.48	9.81 a	0.04
0	5	1.05 ab	0.02	1.14 a	0.02	0.37 ab	0.13	1.04 abc	0.86	2.10 a	0.05	2.91 ab	0.33	−0.03 ab	0.01	−0.02 d	0.01	0.06 a	0.08	53.30 ab	2.68	2.46 b	1.75
10	0	5.42 ab	0.43	1.12 a	0.03	0.33 b	0.02	1.59 ab	0.15	1.53 abc	0.14	2.35 ab	0.39	−0.10 bc	0.02	0.42 abc	0.06	−0.01 abc	0.00	54.65 ab	2.33	10.21 a	1.18
10	2	0.93 b	0.02	1.11 a	0.00	0.43 ab	0.03	1.14 abc	0.59	1.99 ab	0.16	3.01 a	0.12	−0.04 ab	0.01	0.57 a	0.51	0.04 a	0.00	52.75 ab	2.47	3.70 b	2.44
10	5	1.07 ab	0.21	1.12 a	0.05	0.32 b	0.02	0.76 bc	0.02	0.37 d	0.00	1.34 b	1.25	−0.04 ab	0.04	0.12 bcd	0.01	−0.08 c	0.05	52.00 ab	1.41	3.40 b	3.08
20	0	6.15 a	1.24	1.18 a	0.16	0.34 ab	0.00	0.62 c	0.09	1.60 abc	0.09	2.46 ab	0.26	−0.11 c	0.04	0.07 cd	0.01	0.05 a	0.00	54.00 ab	1.41	9.92 a	1.52
20	2	4.59 ab	5.55	1.17 a	0.04	0.46 a	0.04	0.76 c	0.30	1.10 c	0.78	2.70 ab	0.04	−0.06 abc	0.00	0.09 cd	0.06	0.02 ab	0.03	51.55 ab	2.61	3.73 b	2.33
20	5	0.96 b	0.19	1.14 a	0.00	0.37 ab	0.00	0.69 c	0.06	0.33 d	0.09	1.44 ab	1.59	−0.03 a	0.00	0.11 bcd	0.00	0.00 abc	0.01	50.7 b	1.41	3.75 b	3.13

[1] Different letters for a given dependent variable denote significant differences (α = 0.05) across all treatment conditions. [2] SD is standard deviation. [3] α shows protein breakdown, β shows gelatinization and γ shows cooking stability rate, Abs is Absorption.

Table 9. Main effects on RVA operational parameters [1].

Effects (% Substitution)	Peak 1 (cP) Mean	SD[2]	Trough 1 (Nm) Mean	SD	Break Down Mean	SD	Final Visc[3] (cP) Mean	SD	Setback (cP) Mean	SD	Peak Time (min) Mean	SD	Pasting Temp (°C) Mean	SD
DDGS (%)														
0	1532.0 a	309.27	839.33 a	360.90	692.66 a	91.41	3243.83 a	1968.38	2404.50 a	1648.26	5.93 a	0.56	86.13 a	10.18
10	1208.17 b	133.12	613.33 b	181.97	594.83 b	56.79	2152.50 b	959.03	1539.17 b	785.28	5.53 b	0.29	85.21 a	8.59
20	1017.50 b	72.70	548.83 b	23.49	468.66 c	65.91	1561.00 b	322.52	1012.17 b	317.90	5.33 b	0.04	88.36 a	1.77
SSL (%)														
0	1121.67 b	71.48	474.33 b	59.53	647.33 a	98.98	993.50 b	250.70	519.16 b	233.97	5.25 b	0.05	79.15 b	8.99
2	1233.33 ab	221.02	661.33 ab	163.97	572.00 b	106.28	2676.50 a	1089.27	2015.17 a	935.38	5.65 a	0.35	89.30 a	0.75
5	1402.67 a	423.11	865.83 a	311.94	536.83 b	132.10	3287.33 a	1418.14	2421.50 a	1132.29	5.88 a	0.51	91.25 a	2.02

[1] Different letters for a given dependent variable denote significant differences ($\alpha = 0.05$) across treatment conditions for that independent variable. [2] SD is standard deviation. [3] Visc is viscosity. RVA is Rapid Visco Analyzer.

Table 10. Treatment effects on RVA operational parameters [1].

DDGS (%)	SSL (%)	Peak 1(cP) Mean	SD[2]	Trough 1 (Nm) Mean	SD	Break Down Mean	SD	Final Visc[3] (cP) Mean	SD	Setback (cP) Mean	SD	Peak Time (min) Mean	SD	Pasting Temp (°C) Mean	SD
0	0	1205.50 d	16.26	455.50 c	0.70	750.00 a	15.55	802.50 f	427.79	347.00 e	428.50	5.26 e	0.00	75.00 b	11.31
0	2	1501.50 b	28.99	842.00 b	171.11	659.50 ab	142.12	3965.00 b	486.48	3123.00 a	315.36	6.06 b	0.18	89.65 a	0.91
0	5	1889.00 a	84.85	1220.50 a	191.62	668.50 ab	106.77	4964.00 a	272.94	3743.50 a	464.56	6.46 a	0.28	93.75 a	1.06
10	0	1082.00 ef	32.52	422.00 c	19.79	660.00 ab	12.72	1024.00 f	46.66	602.00 ed	26.87	5.19 e	0.00	76.12 a	10.92
10	2	1172.00 ed	11.31	595.00 c	57.98	577.00 bc	46.66	2363.00 d	523.25	1768.00 cb	465.27	5.56 bc	0.14	89.50 a	1.06
10	5	1370.50 c	16.26	823.00 b	2.82	547.50 bc	13.43	3070.50 c	3070.50	2247.50 b	5.00	5.83 cd	37.47	90.02 a	0.04
20	0	1077.50 ef	55.86	545.50 c	31.81	532.00 bcd	24.04	1154.00 ef	52.32	608.50 ed	20.50	5.29 ed	0.04	86.32 ab	1.16
20	2	1026.50 fg	79.90	547.00 c	39.59	479.50 cd	40.30	1701.50 ed	61.51	1154.50 cd	21.92	5.33 ed	0.00	88.77 a	0.03
20	5	948.50 g	6.36	554.00 c	9.89	394.50 d	16.26	1827.50 d	27.57	1273.50 c	37.47	5.36 ed	0.04	0.04 a	0.04

[1] Different letters for a given dependent variable denote significant differences ($\alpha = 0.05$) across treatment conditions for that independent variable. [2] SD is standard deviation. [3] Visc is viscosity. RVA is Rapid Visco Analyzer.

Table 11. Main effects on quality evaluation parameters [1].

Effect	Uniformity		Size		Thickness		Softness		Color	
	Mean	SD [2]	Mean	SD	Mean	SD	Mean	SD	Mean	SD
DDGS (%)										
0	7.96 b	0.66	7.30 a	0.74	7.10 a	0.99	8.36 a	0.99	8.30 a	0.74
10	8.60 a	0.77	5.66 b	1.34	6.23 b	1.30	4.86 c	1.27	7.70 b	0.53
20	8.50 a	0.93	7.20 a	0.84	6.90 a	1.24	6.93 b	1.25	6.93 c	0.78
SSL (%)										
0	8.56 a	1.04	6.50 b	1.50	6.96 a	0.96	6.53 b	1.96	7.50 a	0.820
2	8.40 ab	0.77	7.26 a	0.04	6.44 a	1.66	6.33 b	1.62	7.63 a	1.88
5	8.10 b	0.60	6.40 b	1.10	6.80 a	1.39	7.26 a	1.91	7.80 a	0.55

[1] Different letters for a given dependent variable denote significant differences ($\alpha = 0.05$) across treatment conditions for that independent variable. [2] SD is standard deviation.

Table 12. Treatment effects on quality evaluation parameters [1].

DDGS (%)	SSL (%)	Uniformity		Size		Thickness		Softness		Color	
		Mean	SD [2]	Mean	SD	Mean	SD	Mean	SD	Mean	SD
0	0	7.70 d	0.82	7.10 ab	0.87	7.00 b	0.94	8.50 ab	0.89	8.10 b	0.73
0	0	7.90 d	0.56	7.70 a	0.82	7.50 ab	1.26	7.80 bc	0.918	8.80 a	0.78
0	0	8.30 bcd	0.48	7.10 ab	0.31	6.80 bc	0.63	8.80 a	1.03	8.00 bc	0.41
10	2	9.10 a	0.87	4.90 c	1.97	7.00 b	1.05	4.20 e	0.42	7.70 bc	0.48
10	2	8.60 abc	0.51	6.70 b	1.05	6.10 cd	0.87	5.40 d	1.64	7.50 c	0.70
10	2	8.10 cd	0.56	5.4 c	1.17	5.60 d	1.57	5.00 ed	1.24	7.90 bc	0.31
20	5	8.90 ab	0.87	7.50 ab	0.84	6.90 bc	0.94	6.90 c	0.99	6.70 d	0.48
20	5	8.70 abc	0.94	7.40 ab	0.69	5.80 d	0.42	5.90 d	1.19	6.60 d	0.84
20	5	7.90 d	0.73	6.70 b	0.82	8.00 a	0.47	8.00 ab	0.47	7.50 c	0.70

[1] Different letters for a given dependent variable denote significant differences ($\alpha = 0.05$) across treatment conditions for that independent variable. [2] SD is standard deviation.

3.1. Chemical Properties

3.1.1. Protein

The protein content of bread can be influenced by the Maillard reaction and also the aggregation, which can happen due to the dehydration of the surface due to high temperature during baking. In this study, the results from the main effect of DDGS showed that the highest content of protein was obtained in the bread substituted with 20% DDGS, which was expected since DDGS is a rich source of protein. As for SSL, the highest protein value was for bread with 0% SSL, while the lowest was for the bread made with 5% SSL. Thus, in the treatment effects, the highest value for protein content was the bread made with 0% SSL and 20% DDGS; a significant difference was observed between breads made with this treatment and control. This result is in accordance with a study by Reddy et al. [16], which showed that addition of DDG in the production of muffins resulted in higher protein content compared to controls. Another study showed that the addition of DDGS to chocolate chip cookies resulted in an increased protein content by 30% [17]. The protein content of the bread can be the direct reflection of fermentation in dough, because fermentation is an important step for the protein solubilization [18].

3.1.2. Fiber

As expected, the highest value of fiber was in the bread made with 20% DDGS and 0% of SSL, with a significant difference from the control bread. Up to 5% addition of SSL reduced fiber in the bread. DDGS is a very good source of dietary fiber, which can increase the nutritional value and rheological properties of bread. Although fiber can increase the water absorption through hydrogen binding which is due to the hydroxyl groups in the fiber structure, it can reduce bread volume and affect the texture as well [19]. Since the non-enzymatic browning among peptides can produce fiber-like substances, fiber content of Barbari can increase as a result of baking [20]. Our result is consistent with other studies which also showed that incorporation of DDGS in other products had similar results for fiber content. For instance, addition of corn distillers (CDS) into spaghetti resulted in enhancement of fiber content up to 12–14% in comparison with the control sample [8]. In another study, it was shown that addition of DDGS in chocolate chip cookies and banana bread increased the fiber content of final products [17].

3.1.3. Fat

Addition of DDGS at level of 20% resulted in the highest value of fat for the DDGS main effect, and addition of 5% SSL resulted in the highest fat content for the SSL main effect. For the treatment effect combinations, addition of 20% DDGS and 5% SSL, resulted in a significantly higher value of fat content compared to the control bread. According to Rasco et al. [17], this increase can be due to the high fat content of DDGS. Addition of SSL to the bread formulation can help in evenly distribution of lipids, and generation of fatty acid-amylose complex which may remain in the starch granules [21]. Similar results can be observed in other studies. For example, in a study by Reddy et al. [16], the amount of lipid in soft wheat DDG increased up to 1.4–2.4 times when compared to whole grain wheat prior to fermentation; however, incorporation of 10% DDG into wheat muffin did not change the fat content significantly.

3.1.4. Moisture

In general, Barbari bread has the highest moisture content among Iranian breads [20]. However, in this study, the highest moisture content as a main effect of SSL was found in the bread without SSL, and for the main effect of DDGS, the highest value was for the bread made without any DDGS. Although in a study, it was shown that addition of DDGS to baked products resulted in an increase in the water absorption [22], in our study, no significant differences were seen between different treatments in moisture content, as they had almost the same amount of moisture.

3.2. Physical Properties

3.2.1. Thickness of Bread

Addition of 20% DDGS incorporated with 2% SSL resulted in the highest value of thickness, likely due to the use of SSL in the bread. The lowest value was for control, which was significantly lower compared to bread made with 20% DDGS and 2% SSL. Thickness can be affected by both the fiber content and the use of hydrocolloids. Dietary fiber, in general, reduces the volume of bread. In a study by Park et al. [23], incorporation of fiber into bread resulted in decrease of volume by 5–15% due to the poor gas retention in bread. Also, another study showed that supplementation of 15 and 25% DDGS reduced the average thickness of cookies [24]. On the other hand, using hydrocolloids in bread formulation, such as SSL, can improve the formations of gluten networks and increase the volume of the bread.

3.2.2. Texture

Texture has a direct impact on consumer acceptability and it has other effects through releasing flavor and its influence on appearance [9]. In this study, two textural attributes of final bread samples were measured; firmness and extensibility. The highest value of extensibility was found in bread made with 0% and 2% SSL, and increasing the value of SSL up to 5% resulted in the lowest amount of extensibility. Also, addition of DDGS up to 20% resulted in decreased extensibility which occurred due to the low amount of gluten content in bread. Thus, the highest value of extensibility was measured in the bread made with 0% DDGS and 2% SSL, and the lowest extensibility value was for the bread made with 20% DDGS and 5% SSL which were significantly different than each other. The most important factor which gives extensibility to bread is the gluten and SSL can help strengthen the gluten network and retain gas which is produced by yeast in the dough [21].

One important factor which can result in an increased firmness of bread is the amount of fiber in the flour. In one study, bread with 2% fiber-supplementation had significantly firmer crumb compared to control bread. This was due to the thickening of walls surrounding the air bubbles in the crumb [19]. The highest amount of firmness, for addition of DDGS was the bread made with 10% DDGS, but it was not significantly different from the 20% DDGS. As for the addition and main effect of SSL, the highest amount of firmness was determined in the bread made with 5% SSL. As for the treatment effect, the treatment with 10% DDGS and 5% SSL had the highest amount of firmness and was significantly different from the control bread. However, in a study by Marco and Rosell [25], addition of hydroxypropyl methylcellulose (HPMC) in bread resulted in a significant decrease in crumb hardness; but addition of protein increased the hardness. Thus, another factor, which can improve bread firmness, is the protein content. In a study by Ahlborn et al. [26] it was shown that the force required to compress bread was higher for the bread with protein added than standard wheat bread. The protein in bread is gluten, which provides unique functional properties in bread and is responsible for the protein-starch interaction, providing specific viscoelastic properties in bread dough [26]. Thus, with the high content of fiber and protein as well as the effect of SSL on gluten formation, it can be implied that both DDGS and SSL affected the firmness and texture of bread samples.

3.2.3. Water Activity

There were no significant differences in water activity between different treatments, but the treatment with 20% DDGS and 0% SSL was slightly higher compared to the other treatments. In a study by Marco and Rosell [25], addition of HPMC resulted in a significant decrease of moisture due to hydrocolloids retaining water. Water has an important role in the production of bread; it takes part in starch gelatinization, protein denaturation, flavor, and color development. Fiber plays a role in the water absorption; as the amount of fiber increases, the flour-water absorption increases as well [27].

3.2.4. Color

Color is a quality parameter which affects consumer acceptability of bread. The formation of the golden yellow color on Barbari bread is produced because of the Romal which is brushed on the bread. Romal is made from baking soda and wheat flour dissolved in warm water, which leads to the formation of dextrin during baking and finally the golden color. Thus, the type of flour and the browning reaction which are non-enzymatic reactions, play important roles. Since bread contains both reducing sugars and amino groups, when it is heated, caramelization and Maillard reaction may take place at the same time. One study showed that in order for browning reactions to take place, temperature greater than 120 °C and water activity less than 0.6 are required [28]. In this study, L^*, a^* and b^* were determined for bread samples. In terms of DDGS main effect, addition of 20% DDGS had the highest L^* value which is due to the natural dark color of DDGS. For SSL main effect, there were no significant differences between different values of SSL. Overall, bread made with 20% DDGS and 2% SSL had the highest value of L^* which was significantly different than breads made with 0% DDGS, 2 and 5% of SSL. This was due to the incorporation of SSL in the flour matrix and Maillard reactions during baking which contributes to the golden and brown color of breads. The a^* value was highest for the bread made with 10% DDGS, so it was more red than all others; however, there were really no significant differences for different amounts of DDGS in formulations. As for SSL, the highest value was for no addition of SSL. The treatment effect showed the highest value was for the control bread, but there were no significant differences between different treatments. The highest b^* value was in the bread made with 20% DDGS and 5% SSL, which resulted in yellower bread, while the b^* value for the control bread and breads made with 20% DDGS, 2 and 5% SSL had no significant differences. In a study by Rasco et al. [17], it was shown that 30% blends of all-purpose flour and DDGS had a darker color, more red and more yellow than control blends. Color also can be affected by the raw ingredients, especially flour. DDGS is an ingredient which can have a great impact on the final product's color. Different bakery products have been tested and made with DDGS. For example, adding 30% DDGS to muffins made them darker in comparison to control muffins, and adding up to 20% DDGS resulted in a darker color in hush puppies [29].

3.2.5. Quality

In this study, Barbari breads were also tested for quality attributes such as uniformity, thickness, softness, size and color by the test panelists. Bread made with 10% DDGS and 0% SSL had the highest uniformity, and was significantly different than control bread. Other studies show that adding up to 10% DDG to wheat muffins changed neither appearance nor texture; however, a grainy texture in muffins supplemented with 20% DDG was found [16]. In another study, adding 30% DDG to muffins made the muffins dry and more irregular in cell distribution compared to the control [29]. As for the size, the highest amount was related to the bread made with 0% of DDGS and 2% of SSL but it was not significantly different than the control; the lowest was determined for the bread made with 10% DDGS and 0% SSL. In evaluation of bread thickness, bread made with 20% DDGS and 5% SSL had the highest amount, while the one made with 10% DDGS and 5% SSL had the lowest thickness.

The results for softness of samples showed that bread made with 0% DDGS and 5% SSL was softest. Addition of emulsifiers to baked products can lead to crumb softening because of the interaction of emulsifiers with the starch. Also, by preventing amylose and amylopectin retrogradation, they can inhibit bread staling [27]. Finally, in evaluation of color (Figure 1), bread made with 0% DDGS and 2% SSL had a better color which was significantly different than the control (0% DDGS and 0% SSL). In general, as DDGS increased, color decreased.

DDGS (%)   0   10   20

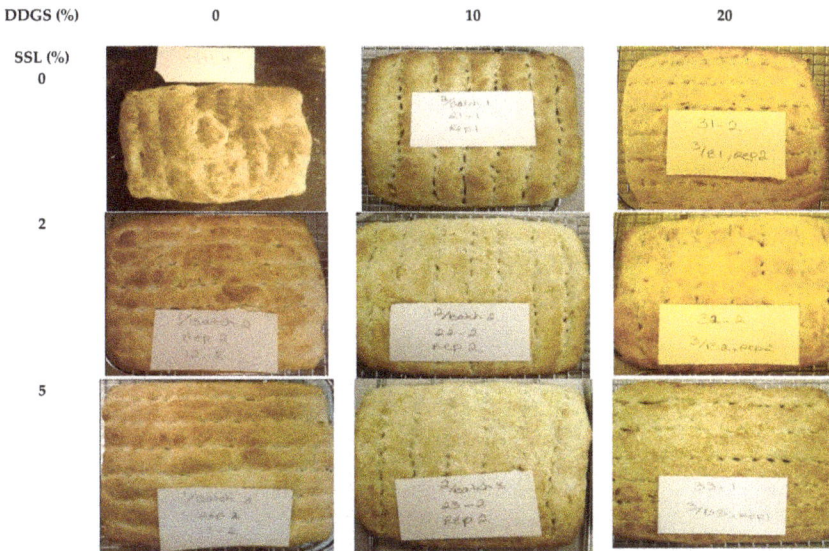

Figure 1. Final bread products baked with different blends of distillers dried grains with solubles (DDGS) and sodium stearoyl-2-lactate (SSL).

3.3. Mixolab Results

As shown in Table 8, different Mixolab parameters were measured in this study. Development time (C1) was a maximum value when 10% DDGS and 5% SSL were used, this part of the curve is where water is being added to the flour and the starch granules and proteins start to absorb water, making them swelled; C1 values were close to each other, and there were no significant differences between different levels of DDGS and SSL which indicates resistance of the dough in all treatments. The measurement of minimum torque (C2), which shows the protein behavior, also didn't show significant differences between different treatment effects. This part of the curve shows the protein behavior and how strong the protein can be in the dough network. Since addition of DDGS added to the protein value of the dough, it is reasonable to see insignificant differences in the C2 values. As for peak torque (C3), the highest value was for control sample, which indicates better quality of starch and shows a decrease as DDGS increased. The cooking stability (C4), which is the indication of amylastic activity, had the highest value for 0% DDGS and 5% SSL, and the lowest value was measured when 20% DDGS and 5% SSL were used, with significant differences between the two treatments. The set back (C5) had the highest value for 10% DDGS and 2% SSL, and it had a significant difference in treatment with lowest value of C5 in which 10% DDGS and 5% SSL were used; in this region retrogradation of starch happens, which caused an increase in the consistency as the temperature was decreasing.

Other parameters determined for Mixolab measurement were α, with the highest value for treatment with 20% DDGS and 5% SSL, which indicated that addition of DDGS to the wheat flour can help improving the protein network of the dough, as α shows the protein break down in the dough matrix. β, which is a good indicator of starch gelatinization, had the highest value for the treatment with 10% DDGS and 2% SSL, and lowest value for 0% of DDGS and 5% SSL, with significant differences between them; and γ, which shows the rate at which cooking stability is reached, had the highest value in the treatment with 0% DDGS and 5% SSL. Water absorption in the developing dough had its highest value for the treatment of 0% DDGS and 2% SSL, with significant difference from that of control dough. Also, the stability time had its highest value for the control sample, but the lowest value for the incorporation of 5% SSL and 0% DDGS. Stronger wheat flours have the ability to absorb and retain more water compared to weak flours (flours with lower protein content). Higher development time indicates stronger flour (flours with higher protein content) [12]. These results are in accordance with other studies; for instance, using an amylograph, it was determined that the addition of SSL to defatted soybean meal and defatted sesame meal caused slight delay of on-set of starch gelatinization and the flour with higher SSL had higher water absorption [30]. In another study, addition of surfactants decreased water absorption of flour by about 0.4–1.2%, and the dough development time decreased by 0.5 min with SSL and DATEM (mono and diacetyltartaric acid) combination [31]. In the study by Tsen et al. [9], replacement of flour by 10 to 20% DDG reduced dough development time.

3.4. Rapid Visco Analyzer (RVA) Results

RVA can be used to measure pasting properties. The RVA pasting curve can detect differences in flour viscosity with a small amount of sample and in a short period of time [32]. Because in our study, SSL was used, peak viscosity, pasting temperature and setback measured by the RVA can be useful in predicting the dough behavior during baking. Wheat chemistry is complex, involving interactions between gliadin and glutenin as well as starch and gluten network. Interactions between gliadin and glutenin are responsible for the initial RVA viscosity [33], and these interactions can affect bread quality. In our study, the highest peak value occurred for the treatment in which 0% DDGS and 5% SSL was used, which was significantly different from the sample made with 20% DDGS and 5% SSL. When starch granules start swelling, viscosity of the paste will increase until a maximum viscosity is reached at "peak viscosity" which is the reflection of water binding capacity of the starch [34]. The trough value was the highest when 0% of DDGS and 5% of SSL were used compared to the control sample which had the lowest value. When 20% DDGS and 5% of SSL were used, the break down had its lowest value compared to the control sample with highest value. In one study, the behavior of wheat flour subjected to shear stress was affected by adding hydrocolloids, causing an increase in water absorption and dough development time [12]. The final viscosity had the highest value with incorporation of 5% SSL and 0% DDGS which were significantly different from the control. As for the setback time, the highest value was measured when 0% DDGS were used and 5% SSL. The control sample had the lowest peak time, and the highest peak time value occurred when 0% DDGS and 5% SSL were used, with significant differences between these two treatments. The last parameter measured was pasting temperature, with the highest value when 0% DDGS and 5% SSL were used. This was expected because of the role of SSL in interaction between starch and gluten network. Changes in the gluten network, such as conformational modifications and loss of hydrogen bonds can decrease RVA viscosity during heating [33]. On the other hand, higher swelling capacity of starch will result in a higher peak viscosity, as the ability of the starch to withstand high temperatures and shear stresses is important [35].

4. Conclusions

Flat breads made from various types of flours, especially wheat flour, are the oldest type of food. Certain deficiencies and nutritional problems exist with cereal-based products, particularly breads. Because of the high demand for flat breads in most countries, fortification of breads can help providing additional nutrients to consumers. DDGS is a good source of protein as well as fiber and can be an inexpensive source for fortification of flat breads. In this study, three levels (0%, 10% and 20%) of DDGS were added to the wheat flour to study the changes in the physical and chemical properties of the final bread products. In addition to the DDGS, SSL was also added to the formulation at three levels (0%, 2% and 5%). The resulting breads were measured for their physical and chemical properties. Overall, the results of this study showed that addition of DDGS can increase fiber and protein values of Barbari bread significantly, while addition of 5% SSL can lead to a softer texture of bread. As the DDGS content increased in the bread formulations, the L^* increased because of the darkness in the DDGS. In addition to the physical and chemical properties of the final products, the rheological properties of the flours were also measured using Mixolab and RVA. The results showed that water absorption of the developing dough had its highest value for the treatment of 0% DDGS and 2% of SSL. For the RVA, highest amount of peak value was for the treatment in which 0% of DDGS and 5% of SSL. The final viscosity had the highest value with incorporation of 5% SSL and 0% of DDGS.

Acknowledgments: The authors would like to thank the USDA, ARS for providing funding, as well as South Dakota State University for use of facilities and equipment.

Author Contributions: K.A.R. and R.G.K conceived and designed the experiments; S.P. performed the experiments; S.P. and K.A.R. analyzed the data; S.P. and K.A.R. wrote the paper; K.A.R. and R.G.K edited the paper.

Conflicts of Interest: The authors declare no conflicts of interest.

References

1. Mckevith, B. Nutritional aspects of Cereals. *Nutr. Bull.* **2004**, *29*, 111–142. [CrossRef]
2. Qarooni, J. *Flat Bread Technology*; Chapman and Hall: New York, NY, USA, 1996.
3. Schilling, M.W.; Battula, V.; Loar, R.E.; Jackson, V.; Kin, S.; Corzo, A. Dietary inclusion level effects of Distillers Dried Grains with Solubles on broiler meat quality. *Poult. Sci.* **2010**, *89*, 752–760. [CrossRef] [PubMed]
4. Bobcock, B.A.; Hayes, D.J.; Lawrence, J.D. *Using Distillers Grain in the US and International Livestock and Poultry Industry*; Midwest Agribusiness Trade Research and Information Center: Ames, IA, USA, 2008.
5. Belyea, R.L.; Rausch, K.D.; Tumbleson, M.E. Composition of corn and distillers dried grains with soluble from dry grind ethanol processing. *Bioresour. Technol.* **2004**, *94*, 293–298. [CrossRef] [PubMed]
6. Wu, Y.V.; Youngs, V.L.; Warner, K.; Bookwalter, G.N. Evaluation of spaghetti supplemented with corn distillers dried grains. *Cereal Chem.* **1987**, *64*, 434–436.
7. Finley, J.W.; Hanamoto, M.M. Milling and baking properties of dried brewer's spent grains. *Cereal Chem.* **1980**, *57*, 166–168.
8. Tsen, C.C.; Eyestone, W.; Weber, J. Evaluation of the quality of cookies supplemented with distillers dried grain flour. *J. Food Sci.* **1982**, *47*, 648–685. [CrossRef]
9. Tsen, C.C.; Weber, J.L.; Eyestone, W. Evaluation of DDG flour as a bread ingredient. *Cereal Chem.* **1983**, *60*, 295–297.
10. Liu, S.X.; Singh, M.; Inglett, G. Effect of incorporation of Distillers Dried Grain with Solubles (DDGS) on quality of cornbread. *Food Sci. Technol.* **2010**, *44*, 713–718. [CrossRef]
11. Caonkar, A.G.; Mcpherson, A. *Ingredient Interactions Effect on Food Quality*; Taylor and Francis Group: Boca Raton, FL, USA, 2006.
12. Rosell, C.M.; Collar, C.; Haros, M. Assessment of hydrocolloid effects on the thermo-mechanical properties of wheat using the Mixolab. *Food Hydrocoll.* **2007**, *21*, 452–462. [CrossRef]
13. Hebeda, R.E.; Zobel, H.F. *Baked Goods Freshness Technology, Evaluation and Inhibition of Staling*; CRC Press: New York, NY, USA; Marcel Dekker: New York, NY, USA, 1996.
14. AACC International. *Approved Methods of Analysis*, 11th ed.; Methods 08-03, 30-25, 44-19 and 46-30; AACCI: St. Paul, MN, USA, 2010.
15. AOAC International. *Official Methods of Analysis of the Association of Analytical Chemists*, 17th ed.; Method 920.39; The Association: Gaithersburg, MA, USA, 2003.
16. Reddy, N.R.; Pierson, M.D.; Cooler, F.W. Supplementation of wheat muffins with dried distillers grain flour. *J. Food Qual.* **1986**, *9*, 243–249. [CrossRef]
17. Rasco, B.A.; Dong, F.M.; Hashisaka, A.E.; Gazzaz, S.S.; Downey, S.E.; Buenarentura, M.S. Chemical composition of distillers dried grains with solubles from soft white wheat, hard red wheat and corn. *J. Food Sci.* **1987**, *52*, 236–237. [CrossRef]
18. Horszwald, A.; Troszynska, A.; Delcastillo, D.; Zielinski, H. Protein and sensorial properties of rye breads. *Eur. Food Res. Technol.* **2009**, *229*, 875–886. [CrossRef]
19. Gomez, M.; Ronda, F.; Banco, C.A.; Caballero, P.A.; Apestegulia, A. Effect of dietary fiber on dough rheology and bread quality. *Eur. Food Res. Technol.* **2003**, *216*, 51–56. [CrossRef]
20. Faridi, H.A.; Finney, P.L.; Rubenthaler, G.L.; Hubbard, J.D. Functional (bread making) and compositional characteristics of Iranian flat breads. *J. Food Sci.* **1982**, *47*, 926–929. [CrossRef]

21. Ghanbari, M.; Shahedi, M. Effect of shortening and SSL on the staling of Barbari bread. *J. Sci. Technol. Agric. Nat. Resour.* **2008**, *43*, 382–390.
22. Abbott, J.; O'Palka, J.; McGuire, C.F. DDGS: Particle size effect on volume and acceptability of baked products. *J. Food Sci.* **1991**, *5*, 1323–1326. [CrossRef]
23. Park, H.; Seib, P.A.; Chung, O.K. Fortifying bread with a mixture of wheat fiber and psyllium husk fiber plus three antioxidants. *Cereal Chem.* **1997**, *74*, 207–211. [CrossRef]
24. Sahlstrom, S.; Baevre, A.B.; Brathen, E. Impact of starch properties on hearth bread charactristics. *J. Cereal Sci.* **2002**, *37*, 275–284. [CrossRef]
25. Marco, C.; Rosell, C.M. Breadmaking performance of protein enriched, gluten free breads. *Eur. Food Res. Technol.* **2008**, *227*, 1205–1213. [CrossRef]
26. Ahlborn, G.J.; Pike, O.A.; Hendriz, S.B.; Hess, W.M.; Huber, C.S. Sensory, mechanical and microscopic evaluation of staling in low-protein and gluten free breads. *Cereal Chem.* **2005**, *82*, 328–335. [CrossRef]
27. Ribotta, P.D.; Perez, G.T.; Leon, A.E.; Anon, M.C. Effect of emulsifier and guar gum on micro structural, rheological and baking performance of frozen bread dough. *Food Hydrocoll.* **2004**, *18*, 305–313. [CrossRef]
28. Parlis, E.; Salvadori, V.O. Modeling the browning of bread during baking. *Food Res. Int.* **2009**, *42*, 865–870. [CrossRef]
29. Brochetti, D.; Penfield, M.P. Sensory characteristics of bakery products containing distillers dried grains form corn, barley and rye. *J. Food Qual.* **1989**, *12*, 413–426. [CrossRef]
30. Safdar, M.N.; Naseem, K.; Siddiqui, N.; Amjad, M.; Hameed, T.; Khalil, S. Quality evaluation of different wheat varieties for the production of unleavened flat bread (chapatti). *Pak. J. Nutr.* **2009**, *8*, 1773–1778. [CrossRef]
31. Azizi, M.H.; Rao, G.V. Effect of surfactant gel and gum combination on dough rheological characteristics and quality of bread. *J. Food Qual.* **2004**, *27*, 320–336. [CrossRef]
32. Sabanis, D.; Lebesi, D.; Tzia, C. Effect of dietary fiber enrichment on selected properties of gluten-free bread. *Food Sci. Technol.* **2009**, *42*, 1380–1389.
33. Lagrain, B.; Thewissen, B.G.; Brijs, K.; Delcour, J.A. Mechanism of glidin-glutenin cross-linkage during hydrothermal treatment. *Food Chem.* **2008**, *107*, 753–760. [CrossRef]
34. Farahnaki, A.; Majzoobi, M. Physiochemical properties of partbaked breads. *Int. J. Food Prop.* **2008**, *11*, 186–195. [CrossRef]
35. Ragaee, S.; Abdel-Aal, E.M. Pasting properties of starch and protein in selected cereals and quality of their products. *Food Chem.* **2006**, *95*, 9–18. [CrossRef]

foods

MDPI

Review

Recent Advances in Physical Post-Harvest Treatments for Shelf-Life Extension of Cereal Crops

Marcus Schmidt [1], **Emanuele Zannini** [1] **and Elke K. Arendt** [1,2,*]

[1] School of Food and Nutritional Sciences, University College Cork, Western Road, T12 Y337 Cork, Ireland; marcus.schmidt@ucc.ie (M.S.); e.zannini@ucc.ie (E.Z.)

[2] Alimentary Pharmabotic Centre Microbiome Institute, University College Cork, T12 Y337 Cork, Ireland

* Correspondence: e.arendt@ucc.ie; Tel.: +353-21-490-2064; Fax: +353-21-427-0213

Received: 30 January 2018; Accepted: 21 March 2018; Published: 22 March 2018

Abstract: As a result of the rapidly growing global population and limited agricultural area, sufficient supply of cereals for food and animal feed has become increasingly challenging. Consequently, it is essential to reduce pre- and post-harvest crop losses. Extensive research, featuring several physical treatments, has been conducted to improve cereal post-harvest preservation, leading to increased food safety and sustainability. Various pests can lead to post-harvest losses and grain quality deterioration. Microbial spoilage due to filamentous fungi and bacteria is one of the main reasons for post-harvest crop losses and mycotoxins can induce additional consumer health hazards. In particular, physical treatments have gained popularity making chemical additives unnecessary. Therefore, this review focuses on recent advances in physical treatments with potential applications for microbial post-harvest decontamination of cereals. The treatments discussed in this article were evaluated for their ability to inhibit spoilage microorganisms and degrade mycotoxins without compromising the grain quality. All treatments evaluated in this review have the potential to inhibit grain spoilage microorganisms. However, each method has some drawbacks, making industrial application difficult. Even under optimal processing conditions, it is unlikely that cereals can be decontaminated of all naturally occurring spoilage organisms with a single treatment. Therefore, future research should aim for the development of a combination of treatments to harness their synergistic properties and avoid grain quality deterioration. For the degradation of mycotoxins the same conclusion can be drawn. In addition, future research must investigate the fate of degraded toxins, to assess the toxicity of their respective degradation products.

Keywords: cereal grains; shelf life; spoilage microorganisms; mycotoxins; physical decontamination

---

1. Introduction

At a time of rapid growth in global populations, sufficient nutritional supply to humanity has become increasingly challenging. On the basis of their long tradition as global staples of the human diet and livestock feed, agricultural crops such as cereals will have a key role in satisfying this growing nutritional need. However, global agricultural area is limited, making it difficult to expand cereal production. Considering that approximately 15% of all cereals worldwide are lost due to microbial pests [1], the most sensible approach to combat this issue is to increase both food safety and sustainability to reduce economic losses. Pre- and post-harvest microbial spoilage counts as one of the predominant factors in crop loss all over the world. Various strategies to prevent microbial contamination in the field have been investigated and reviewed by Oerke [2]. However, even the best management practices cannot completely eliminate the risk of contamination. Because of the permanent and ubiquitous presence of microorganisms and fungal spores in the environment, cereals always carry a certain microbial load when harvested. Additionally, climatic conditions, such as temperature and humidity, that are not under human control may be crucial for contamination with moulds [3]. Therefore, appropriate post-harvest crop treatment, before and during storage, is as important as pre-harvest strategies in the prevention of microbial spoilage. Thus, this review is focused exclusively on research regarding post-harvest treatments.

Depending on climatic conditions during growth, grains carry a microbial load with a high diversity of potential spoilage organisms when harvested. In addition, post-harvest contamination during transport is possible. This microbial load consists of bacteria, yeasts, and filamentous fungi belonging to many different genera. The activity of these micro-organisms during storage and, accordingly, the shelf life of the crop is dependent on a range of factors, as illustrated in Figure 1. Amongst the most influential parameters are moisture content and water availability during storage. As a result, grains are usually stored at low moisture contents of 12–13% and a water activity of <0.70 [4]. However, cereals are usually traded as wet weight and thus inefficient drying systems can lead to microbial spoilage during storage. Furthermore, even if dried properly, some xenerophilic species of *Aspergillus* can still develop during storage, resulting in quality deterioration and mycotoxin accumulation [5].

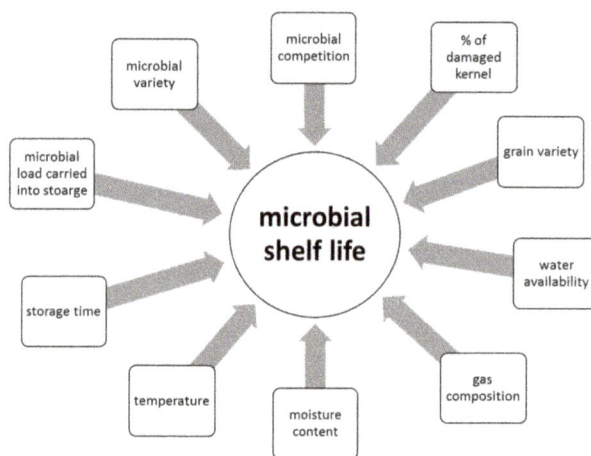

Figure 1. Biotic and abiotic factors influencing the microbial shelf-life of cereal grains during storage.

In addition, conventional storage systems, such as silos, are often cost intensive and inflexible as to volume. In these rigid systems, due to the inappropriate size for the amount of grains stored, the environmental conditions in the headspace cannot be controlled. Thus, it is likely that suitable conditions arise which promote microbial growth and the production of toxic metabolites [6]. As shown in a recent study conducted by Schmidt et al. [7], if the conditions during storage are suitable, a minimal fungal field contamination can rapidly evolve into a serious consumer health hazard.

The biggest microbial threat during storage is displayed by filamentous fungi, mostly belonging to the genera *Aspergillus* and *Penicillium*. This is largely a result of their relative tolerance to low water activities and the production of mycotoxins, which are secondary fungal metabolites with toxic effects on humans and animals. In addition, these fungi induce a loss of nutritional value in grain, produce off-odours, and result in reduced baking and milling quality [4]. Although *Fusarium* spp. are typical field pathogens and unlikely to develop during storage, previously produced and accumulated mycotoxins remain a serious issue during cereal storage and processing [8]. Hence, in addition to the living organisms, a broad variety of mycotoxins must be countered during post-harvest treatments, as the prevention of their production is not always possible. Table 1 shows the most commonly reported mycotoxins of cereal crops and the producing fungal genera. It has been reported that approximately 25% of the global cereal crops, equivalent to over 500 million tons per annum, are contaminated with mycotoxins and thus present a potential consumer health hazard [9]. In contrast to fungal mycelia, mycotoxin-contaminated grains often do not vary visually from clean grains and are therefore difficult to identify and eliminate.

Table 1. Chemical structures of mycotoxins commonly found on cereal grains sorted by the producing fungal species.

Aspergillus spp.	*Aspergillus* spp./*Penicillium* spp.	*Fusarium* spp.
Aflatoxin B$_1$		Deoxynivalenol
Aflatoxin B$_2$	Ochratoxin A	Nivalenol
Aflatoxin G$_1$	Citrinin	Fumonisin B$_1$
Aflatoxin G$_2$		Fumonisin B$_2$
		Zearalenone

Previous studies have established that mycotoxins are primarily located in the husk layers of grains. A study conducted by Vidal et al. [10] showed that wheat bran obtained from a Spanish market contained substantial amounts of various mycotoxins, mainly deoxynivalenol (DON) and zearalenone (ZEA). The authors concluded that the production of whole grain products from these wheat samples would constitute a substantial consumer health hazard. However, since wheat bran is rich in dietary fibre, there is a high interest in exploiting its application in food and feed products because of its potential health-promoting properties. In addition to these *Fusarium* toxins, different toxins produced by storage fungi, namely aflatoxins (AFs), ochratoxin A (OTA), and citrinin are commonly found as contaminants in cereals (Table 1). Finally, bacterial spoilage organisms must also be considered to ensure sufficient grain shelf life. Although most bacteria are unlikely to grow under conditions commonly applied to grain storage, their presence can result in significant quality deterioration during subsequent processing or in the final product [11].

In recent years, the consumer desire for more natural, less processed foods with fewer or no chemical additives has increased enormously; however, the requirement for the maintenance of the highest safety standards remains. This increases the emphasis on physical and microbiological treatments to control post-harvest microbial spoilage in cereals [12]. In addition, the application of both physical and microbiological decontamination methods into necessary food processing procedures while simultaneously enabling a "clean label" has attracted increasing interest from both the industry and researchers. Various approaches for bio-preservation, particularly the use of antifungal LAB (lactic acid bacteria), have been investigated and reviewed elsewhere [5,13,14]. In contrast, this article focuses on the recent advances in novel physical decontamination methods.

Physical grain treatments, including dry and wet heat, ionizing and non-ionizing irradiation, high hydrostatic pressure and modified atmosphere packaging, were critically reviewed as to their suitability to eliminate viable forms and spores of both food spoilage bacteria and fungi commonly found on cereals. Furthermore, the treatments' potential to remove previously produced mycotoxins as well as the impact on grain quality, viability and technological performance were evaluated based on the existing literature. This allowed for the identification of future research needs as well as possibilities for industrial application of the treatments discussed. It should be mentioned that classical physical treatments, such as milling, sorting, and hulling are not covered in this review, as these methods are well established and industrially applied. Therefore, they largely have not been the subject of recent research.

2. Modified Atmosphere Packaging (MAP)

Modified atmospheres have been investigated for the storage of cereals intended for food and feed. While fungi responsible for grain deterioration during storage often are considered obligate aerobes, many are microaerophilic. Hence, they can develop in niches where other species cannot grow and therefore can dominate in grain ecosystems. In many cases, the oxygen level must be <0.14% and the carbon dioxide level >50% to achieve significant growth inhibition [15,16]. Certain species, such as *P. roqueforti* Thom and some *Aspergillus* spp., can grow and infect grains even at >80% carbon dioxide, if at least 4% oxygen is present. In addition, post-harvest systems also use (oxygen-free) nitrogen to prevent grain deterioration [17]. However, it must be noted that results obtained in different sample systems are very difficult to compare, as the tolerance to low oxygen and high carbon dioxide levels is highly dependent on the sample matrix and water availability. Accordingly, low water availability has been reported to increase the sensitivity of microorganisms to the modified atmosphere. While MAP is used to control microbial spoilage and insect pests in moist grains during storage, different threats require different treatment conditions.

Exposure of various toxigenic fungi to ozone gas (60 µmol/mol) for up to 120 min resulted in significantly reduced growth and spore germination for *Fusarium graminearum*, *F. verticillioides* (Sacc.) Nirenberg, *Penicillium citrinum* Thom, C., *Aspergillus parasiticus* Speare, and *A. flavus* Link. However, the efficiency of the treatment was strongly dependent on the specific strain. *F. graminearum* was found to be particularly sensitive, as its growth was totally inhibited after exposure for 40 min, whereas *F. verticillioides* growth after 120 min of exposure was only slightly reduced [18].

However, very little research on modified atmosphere packaging has been conducted in recent years. This can be attributed to various reasons. Firstly, storage conditions under high carbon dioxide and low oxygen levels are difficult to apply to a conventional silo storage system. Additionally, environmental conditions, such as temperature, moisture content, and water availability determine the gas composition required to achieve sufficient microbial inhibition. The biggest potential drawback of this technology is that the microorganisms are not killed and therefore can induce product spoilage during subsequent processing. In addition, the removal of mycotoxins produced in field or the inactivation of microbial enzymes is not possible. In addition, microbial inhibition is substantially dependent on the fungal or bacterial strain. As cereals after harvest carry a wide range of microorganisms, it also appears very difficult to predict the success on a specific crop. Therefore, recent research to prevent post-harvest microbial spoilage has shifted towards novel and more flexible methods for application.

3. Thermal Treatments

Thermal treatments, including pasteurisation and sterilisation techniques, are the most regularly used treatment for the inactivation of microorganisms and enzymes in the food industry [19]. For post-harvest application of heat treatments various approaches, such as hot water dips, hot dry air, or superheated steam, are possible using different time-temperature combinations [20]. Table 2 summarises heat treatments reported to have been successful for microbial inactivation or toxin degradation and the side effects on the sample.

Table 2. The effects of different heat (wet and dry) treatments on microbial load and mycotoxin content in various sample matrices.

Target Organism/Toxin	Treatment	Sample Matrix	Technological Impact	References
Natural microbial load	Dry air 9 day/100 °C	Various cereals	No impact	[21,22]
Natural microbial load	Steam 60 min/82 °C	corn	No impact	[23]
Natural microbial load	Steam 210–250 °C/15 s	Wheat, barley, rye	Not investigated	[24]
F. graminearum	dry air 15 day/60 °C; 5 day/70 °C; 2 day/80 °C	Wheat	No impact	[25]
F. graminearum	Dry air 5 day/90 °C	Wheat	Reduced seed viability	[25]
F. graminearum	Dry air 21 day/60 °C; 9 day/70 °C; 5 day/80 °C	Barley	Reduced viability for 9 day/70 deg; 5 day/80 deg	[25]
Aspergillus spp., *Penicillium* spp., *Fusarium* spp., *E. coli*, *L. Monocytogenes*, *Salmonella* spp.	Steam 170–200 °C/<60 s	Various cereals	No impact	[26–28]
Geobacillus stearothermophilus spores	Steam 20 min/160 °C	Dried spore pellet-sand mixture	Not investigated	[29]
DON (50% reduction)	Steam 6 min/185 °C	wheat	Not investigated	[29]

3.1. Dry Heat Treatments

Described as an alternative to chemical grain disinfection, several applications of dry heat in the form of hot air treatments were studied for the ability to control fungal spoilage without compromising the kernels' viability [25]. Contradictory results have been reported as to the effects of dry heat treatments on the cereal's sensory quality and nutritional value. Several authors reported no impact on technological performance and enzymatic activities after nine days at up to 100 °C [21,22]. In contrast, Gilbert et al. [25] found a significant loss in seed viability after treatment for five days at 90 °C.

For inhibition of various fungal strains on wheat grains, time-temperature combinations of 15 day/60 °C, 5 day/70 °C and 2 day/80 °C, respectively, were found to inhibit fungal growth and spore germination completely without compromising the grain's viability [25,30]. When treating barley at the same temperatures, the fungi were completely inhibited after 21 days, 9 days, and 5 days, respectively. In contrast to wheat, after nine days at 70 °C, the germination capacity of barley was substantially reduced.

Although bacterial spoilage organisms, such as *Bacillus* spp., are unlikely to cause grain quality deterioration during storage, they have to be inactivated prior to grain processing to avoid subsequent product spoilage [11]. However, no studies specifically investigating the heat inactivation of grain spoilage bacteria post-harvest are available to date. In general, bacteria are reported to be less heat stable than fungal mycelia and it can therefore be assumed that they are also inactivated with the abovementioned antifungal treatments [31].

Thus, dry heat treatment prior to grain storage shows potential for the prevention of post-harvest microbial spoilage based on the heat inactivation of the microbes and the reduction of the grain's moisture content and water availability, which further supports microbial inhibition [32].

Depending on the processing conditions and sample matrix (i.e., the presence or absence of yeast), reports suggest a substantial reduction in the occurrence of mycotoxins during heating. However, cereal flours currently are not always heat treated during processing and this approach is very inconsistent. Therefore, an efficient pre-storage treatment to degrade mycotoxins is essential [33]. In general, conventional heat treatment is not suitable for detoxifying crops contaminated with mycotoxins. Commonly found mycotoxins, such as aflatoxins, trichothecenes, OTA, and others reportedly possess great heat stability (>300 °C) and thus cannot be degraded by dry heat without seriously damaging the treated cereal. To date, no studies reporting the total removal of mycotoxins by dry heat treatments in vitro or in situ are available. In addition, the reported partial degradation of different toxins by dry heat was always found to compromise the grain quality and viability. This indicates that dry heat can serve as a sufficient tool for the degradation of mycotoxins only in combination with other treatments [34]. In addition, the thermal degradation of mycotoxin into intermediate constituent products of unknown identity and toxicity must also be considered [35,36].

Available literature suggests that, depending on the target microorganism and cereal substrate, dry heat treatments have the potential to decrease the microbial load without compromising the grain quality. However, such treatments consume considerable energy and time, lasting up to several days. Furthermore, the conditions must be adjusted carefully to achieve satisfying results for both microbial decontamination and grain quality. In addition, the sole use of dry heat is unsuitable for killing spores of heat-resistant bacteria, as the applicable temperatures are too low and do not kill the grains. Hence, bacterial spores enter a dormant state and can germinate once suitable conditions return, particularly during cereal processing [26]. In terms of mycotoxin degradation, dry heat cannot serve as a sole treatment to efficiently remove mycotoxins produced in-field. The temperatures required are not feasible to maintain the grain's technological performance; decomposition is always incomplete and the resulting degradation products could display a health risk of yet unknown potential. Finally, the efficiency of hot air treatments is further compromised by the particle size of the treated sample. Thus, it is less efficient for the treatment of whole grains than for treating the milled product.

3.2. Wet Heat Treatments

Compared to conventional hot air treatments, the use of hot steam appears to be a more efficient approach in terms of both time and energy for microbial decontamination and mycotoxin degradation (Table 2). In addition, food spoilage bacteria and fungi were found to be less heat resistant when in conditions with high water availability [37]. Apart from the classic saturated water steam (up to 100 °C), superheated steam (SS, up to 250 °C) has recently gained considerable interest as a rapid, non-destructive, and safe decontamination method [38]. Hence, recent advances and novel applications of SS will be the focus of this section. SS, having higher enthalpy than saturated steam, can quickly transfer heat to the material being processed, resulting in rapid temperature increase in the sample. In addition, SS is reported to contribute to better product quality. The major advantages of using SS for food processing include better product quality (colour, shrinkage, and rehydration characteristics), reduced oxidation losses, and higher energy efficiency [39].

Researchers using SS achieved microbial decontamination from vegetative forms of food spoilage fungi (*Aspergillus* spp., *Penicillium* spp., *Fusarium* spp.) and bacteria, such as *Escherichia coli* O157:H7, *Salmonella Typhimurium*, *Salmonella enteritidis* phage type 30, and *Listeria monocytogenes*, on different cereals. Treatment times of less than 60 s with water steam temperatures of 170 °C–200 °C were reported to be sufficient to reduce the microbial load below the respective limit of detection [26–28]. However, none of those treatments resulted in a significant reduction of sensory or nutritional quality. In contrast, corn treated with conventional steam at 82 °C had to be exposed for 60 min in order to achieve a significant reduction of the microbial load by 4-log units [23]. No results regarding the impact of the treatment on the grain quality were reported. As a result, despite the ability of both approaches to decontaminate cereals, the use of superheated steam is a much more time efficient solution.

Compared to their vegetative forms, fungal and bacterial spore inhibition presents a much bigger challenge. As mentioned above, bacterial endospores can lay dormant and germinate when conditions are favourable during downstream processing. Nonetheless, SS was found to be efficient at permanently killing spores of *Geobacillus stearothermophilus*. However, exposure times of up to 20 min are required to kill spores. This is significantly longer than the inactivation of vegetative microbes. Thus, the biggest reduction in spore viability was detected during the first 5 min of treatment, independent from the processing temperature [29].

In addition, SS was also investigated for the removal of mycotoxins from different cereal matrices. Partial degradation of DON upon SS treatment was reported by Cenkowski et al. [29]. The authors found that increases in steam temperature and exposure time correlated with higher degradation rates for DON. Thus, the highest treatment temperature (185 °C), combined with the longest treatment time (6 min), resulted in the highest reduction of DON, by 50%, which was found to be independent of the steam velocity. Furthermore, the reduction was found to be exclusively due to thermal degradation, rather than solubilisation and water extraction with the steam. However, no investigation of the treatment's impact on the grain's technological performance were undertaken in the study of Cenkowski et al. [29]. Likewise, the dry decomposition temperature of aflatoxins (approximately 270 °C) is known to be significantly reduced under moist conditions [40]. Consequently, SS treatments present a much more efficient and promising approach for AF (aflatoxins) degradation than conventional dry heating.

In conclusion, the application of wet heat and superheated steam were found to be very effective in the decontamination of cereals without compromising the grain's technological quality and performance. However, despite extensive research, future work is still required to optimise the processing parameters which is dependent on the matrix present. The technological impact of treatments long enough to kill spores and degrade mycotoxins requires further investigation. Additionally, the fate of mycotoxins thermally degraded by the SS remains unclear. Thus, the degree of actual detoxification is still in question. Similarly, the possibility of degrading other toxins, such as patulin or bacterial toxins, requires closer investigation.

Finally, it is noteworthy that ultra-superheated steam (USS) has recently attracted considerable research interest. This technique employs temperatures of 400–500 °C. Exposure of different cereals, namely wheat, barley, and rye for 15 s to USS (actual contact temperature 210–250 °C) resulted in total inhibition of spoilage fungi and grain shelf-life extension without notable quality deterioration [24]. However, few studies have investigated the application of USS for the microbial decontamination of food commodities. Thus, the optimal conditions of use and the full potential of this method remain unclear. Further research is needed to clarify the suitability of USS treatments to decontaminate and potentially detoxify cereal grains post-harvest.

4. Ionizing Irradiation

Ionizing irradiation approaches largely are based on short waves of electromagnetic energy that travel at the speed of light. All treatments discussed in this section share the basic properties of electromagnetic radiation. Ionizing radiation treatment using either gamma rays or an electron beam (e-beam) is well established as a rapid, efficient, safe, and environmentally friendly technique for the reduction of food-borne diseases by destroying pathogenic and toxigenic microorganisms [41]. Their biggest advantage is their great penetrating power (inversely related to the frequency [42]), their high efficiency against various food spoilage organisms, and the absence of a rise in temperature in the treated sample [1]. In addition, irradiation treatments for food processing purposes are unconditionally regarded as safe for dosages of up to 10 kGy (1 Gy = 1 Joule of irradiation energy/kg sample matter, with 1 Joule = $1 \frac{kg \cdot m^2}{s^2}$) [43]. Due to the unit of irradiation dosage containing the ratio of energy per time and sample matter, treatments are commonly compared by means of dose and form of application only. Successful treatments with ionizing irradiation against various food spoilage organisms and toxins are summarised in Table 3.

Table 3. The effects of different ionizing irradiation treatments (gamma- and e-beam irradiation) on microbial load and mycotoxin content in various sample matrices.

Target Organism/Toxin	Treatment	Sample Matrix	Technological Impact	References
Natural fungal population	0.75 kGy gamma	millet	none	[44]
Natural microbial load	6 kGy e-beam	chestnuts	No effect on nutritional value	[45,46]
L. monocytogenes	3.3 kGy e-beam (soft electrons)	Alfalfa sprouts	No quality deterioration	[47]
Aspergillus spp., *Alternaria* spp., *Fusarium* spp., *Curvularia* spp., *Helminthosporium* spp.	1.5–3.5 kGy gamma	wheat	Reduced quality for doses >2.5 kGy	[48]
Fusarium spp.	4 kGy gamma	Barley	Reduced quality	[49]
Fusarium spp.	6 kGy gamma	Wheat and maize	Reduced quality	[49]
Aspergillus spp., *Penicillium* spp.	10–15 kGy e-beam	Dry split beans	No quality deterioration (10 kGy)	[50]
Penicillium spp., *Fusarium* spp., *Aspergillus* spp.	1.7–4.8 kGy e-beam	corn	Not investigated	[41]
Fusarium spp. and DON	6–10 kGy e-beam	Barley, malt	Not investigated	[51]
OTA and aflatoxins	15 kGy gamma	Wheat and sesame	Not investigated	[52–54]
OTA	2 kGy gamma	Aqueous solution	-	[55]
DON, ZEN, T-2, FB_1	10 kGy gamma	Soy beans, corn, wheat	Not investigated	[56]
FB_1	7 kGy gamma	Barley, wheat, maize	Not investigated	[49]
aflatoxins	1.5 kGy e-beam	Ground almond flour	Not investigated	[57]

DON: Desoxynivalenol; OTA: Ochratoxin A; ZEN: Zearalenone; T-2: Fusariotoxin T 2; FB_1: Fumonisin B1; -: none.

4.1. Gamma Irradiation

This section discusses the application of irradiation to decontaminate cereals using gamma rays. These electromagnetic waves are usually produced by radioactive cobalt isotopes (^{60}Co). The use of gamma rays for irradiation treatments is characterised by their high penetration energy and short treatment times.

For millet grains exposed to gamma radiation, no significant decrease in fungal incidence or spore germination was reported for radiation doses up to 0.5 kGy. However, at doses of 0.75 kGy or higher, the rate of fungal incidences and spore germination sharply decreased by over 80% and 2-log units, respectively [44]. Salem et al. [48] applied gamma irradiation ranging from 0.5–3.5 kGy to wheat grains prior to storage. A dosage of 1.5 kGy was found to be sufficient for at least a 90% reduction of *Aspergillus* spp., *Alternaria* spp., *Fusarium* spp., *Curvularia* spp., and *Helminthosporium* spp. immediately after the treatment. However, after six months of subsequent storage, the degree of inhibition (compared to the untreated control) was significantly reduced for all species apart from *Fusarium* spp. The authors also reported that higher irradiation doses resulted in higher inhibition rates. In particular, 3.5 kGy, resulting in total inhibition directly after the treatment, also showed total inhibition after six months against all fungi apart from *Aspergillus* spp. (96.4%). In contrast, Aziz et al. [49] reported higher irradiation doses, 4 kGy for barley and 6 kGy for wheat and maize, necessary for total inhibition of *Fusarium* spp. However, analysis of selected physical, chemical, and rheological properties of these grains prior to and after storage showed that, from a technological performance point of view, irradiation dosages higher than 1.5–2.5 kGy were not feasible. This also correlates with findings reported by other researchers [58,59].

Available literature suggests that the radio sensitivity of different fungal species appears to differ significantly depending on the reference consulted. These discrepancies likely result from countless influencing factors which require further research. These factors include the form of fungal contamination (mycelium or spores) and the moisture contents of spores or commodities. Although moist conditions promote fungal growth and spore germination, dry spores are considered more irradiation resistant. Furthermore, the age of spores and the nature of the matrix irradiated can have a significant influence on the radio sensitivity. In addition, the fungal strain and the temperature before, during, and after irradiation influence the treatment's efficiency for each sample.

However, the most recent research on the application of gamma irradiation on cereals has focused on the degradation of mycotoxins in food and feed commodities. Several studies have investigated the degradation of aflatoxins (AFs) and ochratoxin A (OTA) in particular. Deberghes et al. [60] previously reported on the degradation of OTA in an aqueous solution due to gamma irradiation with 2–5 kGy. However, in situ degradation of mycotoxins such as OTA was found to be much more difficult. After gamma irradiation of wheat and sesame seeds using 15 kGy, degradation of OTA, AFB$_1$ (Aflatoxin B1), AFB$_2$, AFG$_1$ (Aflatoxin G1), and AFG$_2$ with reduction rates of 23.9%, 18.2%, 11.0%, 21.1%, and 13.6%, respectively, were found [52,61]. However, in both studies the application of lower irradiation doses showed no substantial reduction of AFs and OTA. Several other authors also reported similar results on various cereal matrices intended for food or feed use [53,54,62–64]. However, it must be noted that none of these studies took the moisture content of the seeds into consideration, which appears to have a critical impact on the radio stability of the toxins. Therefore, the meaningfulness of the results is limited. In another study, soybeans, corn, and wheat, with respective moisture levels of 9, 13, and 17%, showed no substantial reduction in AFs or OTA after irradiation dosages of up to 20 kGy [56]. Therefore, the success of AF degradation due to gamma rays is not dependent solely on the dose; the moisture content of the sample has a major impact on AF degradation. It is proposed that this is primarily due to the radiolysis of water, producing highly active radicals which can degrade AFs to compounds with lower biological activity [40,65]. Similar results were reported for the degradation of OTA in vitro and in situ. OTA showed great radiostability as a dry substance because irradiation with 8 kGy resulted in no noteworthy reduction. In contrast, when in an aqueous solution (50 ng/mL), significant reduction was achieved with just 2 kGy. Similarly, the degradation in moist wheat kernels (16% moisture) was substantially higher than in dry ones (11% moisture) when irradiated with 8 kGy [55].

On the basis of consulted literature, typical *Fusarium* toxins, such as DON, ZEN (Zearalenone), T-2 (Fusariotoxin T 2), and FB$_1$ (Fumonisin B1) appear to be more radio sensitive than AFs and OTA. DON, ZEN, and T-2 toxin in soybeans, corn, and wheat were significantly reduced by irradiation with 10 kGy [56]. Irradiation of barley, wheat, and maize naturally contaminated with FB$_1$ which resulted in a significant reduction (>85%) and total destruction of the toxin after exposure to 5 kGy and 7 kGy, respectively [49].

Nevertheless, irradiation results in degradation products of unknown identity and toxicity must be considered. Wang et al. [65] detected 20 radiolytic products of AFB$_1$ after irradiation in a water/methanol mixture. To date, just seven of these products have been identified and six are considered less toxic than AFB$_1$ due to the missing double bond on the terminal furan ring (Table 1). Given these results, it remains unclear if degradation of mycotoxins is responsible for the detoxification of the sample or if the resulting products are equally as toxic as the original substance. Furthermore, as the irradiation efficiency is highly dependent on water availability, the application in dry food matrices appears to be limited. Although AFs and OTA belong to the most commonly found toxins in cereal crops, the fate of other contaminants, such as DON, NIV (nivalenol), and ZEN requires further research.

In conclusion, gamma irradiation can reduce and potentially fully inhibit fungal and bacterial spoilage of grains during storage thus avoiding the production of mycotoxins. However, the irradiation dosages required for total inhibition of common storage fungi, such as *Aspergillus parasiticus*, are highly dependent on the sample matrix and fungal load. Necessary doses reported a range between 5 and 10 kGy. Unfortunately, such high irradiation dosages cause severe damage to the treated grains and are therefore not feasible from a technological point of view. On the other hand, a simple reduction of the microbial load is not sufficient either, as even minimal fungal contamination can lead to growth during storage. This ultimately results in huge economic losses and a potential consumer health hazard, as demonstrated by the authors in a previous study [7]. Therefore, gamma irradiation alone cannot serve as an efficient tool for cereal grain preservation during storage but could potentially be applied in combination with other treatments. However, for treatment of cereals purposed for animal feed, higher irradiation dosages are allowed. Thus, gamma irradiation may be a promising approach in the decontamination of animal feed.

4.2. Electron Beam Irradiation

The use of e-beam irradiation has several advantages over the use of conventional gamma irradiation. Table 4 summarises the comparison between gamma and e-beam irradiation. The main advantages include faster operation, lower irradiation dosages, and the use of electricity rather than radioactive materials to generate the electron, making the technology more flexible and easier to use. Unfortunately, its penetration power is lower, largely rendering it a tool for surface disinfection [1]. The general use of e-beam irradiation in the food industry, including the technological background and mode of action, was recently reviewed by Freita-Silva et al. [1]. Accordingly, this section will focus exclusively on recent advances in microbial decontamination using e-beam irradiation. With the exception of Europe, e-beam irradiation has been accepted and is widely applied towards the treatments of various food products worldwide, such as fruits and grains. Approximately 81,593 t are treated annually, out of which only 11 t are from Europe [66].

Table 4. Comparison of gamma (^{60}Co) and e-beam irradiation, adopted from Freita-Silva et al. [1].

Parameters	Gamma Irradiation	E-Beam Irradiation
Irradiation Time	Slow	Fast
Doses (kGy)	Higher doses	Lower doses
Source	Radioactive material	Electricity to generate electrons
Flexibility	Inflexible (cannot be turned off)	More flexible (can be turned off)
Penetration	Good penetration	Lower penetration power

Microbial decontamination and mycotoxin degradation in vitro with the potential of application in situ has been previously reported [67]. The irradiation dose necessary for sufficient decontamination depends on the type and species of grains to be treated [68].

When dry split beans were subjected to e-beam irradiation (0, 2.5, 5, 10, 15 kGy) to control storage moulds, irradiation resulted in a dose-dependent decrease in fungal contaminants. High irradiation doses (10 and 15 kGy) resulted in a complete absence of fungi and undetectable levels of aflatoxins B$_1$ and B$_2$. In contrast, un-irradiated beans carried *Aspergillus niger* van Tieghem at the highest level (33–50%), followed by *A. flavus* (14–20%), and *Penicillium chrysogenum* Thom (7–13%). For total inhibition of fungal incidence, irradiation doses of 10 and 15 kGy were necessary. Irradiated split beans (10 kGy) also showed improved shelf life of up to six months without quality deterioration [50]. In contrast, on raw corn under comparable conditions, doses as low as 1.7, 2.5, and 4.8 kGy were found to be sufficient to fully inhibit the growth of *Penicillium* spp., *Fusarium* spp., and *Aspergillus* spp., respectively [41]. Researchers concluded that, with sufficient optimisation of the processing parameters, e-beam irradiation has considerable potential in the microbial decontamination of cereals. For the reduction of the natural microbial load in chestnuts, 6 kGy were found sufficient for decontamination while nutritionally valuable constituents remained unaffected [45,46,69]. Interestingly, e-beam irradiation of *Fusarium* spp.-infected barley with 6–10 kGy showed no significant reduction in fungal incidence and DON contents in the fresh barley. However, the resulting barley malt was reported to have significantly reduced DON content and fungal occurrence [51]. In another study conducted by Stepanik et al. [70], wheat grains and dried distillers grain were irradiated with up to 55 kGy, which resulted in a maximum DON reduction of 17%.

To date, no studies investigating the fate of mycotoxins exposed to e-beam irradiation are available. This is likely due to the lower energy value of this irradiation compared to gamma irradiation (Table 4). Thus, e-beam irradiation is unlikely to create enough energy for mycotoxin degradation. This applies in particular if the toxins are not exclusively located on the grain surface but also in the inner layers. Therefore, the application on ground almond flour with a dose as low as 1.5 kGy was found to be sufficient for total degradation of aflatoxins [57].

In conclusion, e-beam treatment shows potential to reduce the microbial load and content of mycotoxins produced in-field. However, depending on the sample matrix, microbial load, and target organism, the irradiation dose necessary is likely to exceed that generally permitted, i.e., 10 kGy [43], leading to legislative difficulties. Furthermore, the impact of various e-beam treatments on the grain's sensory and nutritional properties have to be investigated further. Finally, difficulties such as the even treatment of the sample surface and the low penetration energy of the e-beam must be considered before any industrial application.

A variation of conventional e-beam irradiation is the use of so-called "soft electrons". The term "soft" refers to the low energy of the electrons fired at the sample. This results in less impact on the sample in terms of sensory and nutritional value. However, this approach only can serve as a surface treatment, as the e-beam does not have sufficient energy to penetrate the deeper layers of the sample. The use of 3.3 kGy applied at soft electrons was reported to successfully eliminate *L. monocytogenes* from alfalfa sprouts [47]. However, due to the low penetration power, a small particle size and an even sample surface are crucial to ensure uniform treatment. Consequently, the irradiation treatment of soybeans was found to be more difficult, as 17 kGy was insufficient to eliminate the natural microbes present on the surface [71]. Thus, because of the difficult applicability, soft electrons show very little potential for industrial post-harvest decontamination of cereal crops but could have potential for flour treatment after milling.

5. Non-Ionizing Irradiation

This section reviews recent advances in non-ionizing irradiation treatments for microbial decontamination and detoxification with possible applications for cereals. Treatments that were successfully applied for the various target organisms or toxins are summarised in Table 5. Because of the non-ionizing character of the treatments discussed here, the impact on the grain quality is likely to be negligible. In addition, consumer acceptance for ionizing irradiation is relatively low due to misinformation [72]. Use of non-ionizing irradiation should increase consumer acceptance of the final products.

Table 5. The effects of different non-ionizing (light, microwave, ultrasound) treatments on microbial load and mycotoxin content in various sample matrices.

Target Organism/Toxin	Treatment	Sample Matrix	Technological Impact	References
Different food spoilage bacteria and fungi	US (ultrasound) > 60 W/cm^2	Aqueous solution	-	[73]
A. parasiticus	Microwave: 900 W, 2.45 GHz, 1–5 min	Aqueous solution	-	[74]
Aspergillus spp. and *Penicillium* spp.	US: 6 min, 60 °C, 20–39 W/cm^2	Culture medium	-	[75]
Aspergillus spp.	51.2 J/g pulsed white light	wheat	15% reduced seed viability	[76]
Aspergillus spp.	Microwave: 120 s, 2450 MHz, 1.25 kW	Cereals and nuts	Not investigated	[77,78]
Bacillus subtilis	1.0 J/cm^2 * 10 pulses light with 200–1100 nm	spices	No quality deterioration	[79]
OTA, OTB (Ochratoxin B), citrinin	Light: 455 nm/470 nm for 5 day	Aqueous solution	-	[80]
Aflatoxins	UV (Ultraviolet)-light: 265 nm for 15–45 min	nuts	Not investigated	[81]
trichothecenes	US > 200 W/cm^2	corn	No quality deterioration	[82]

* Times (sign for multiplication), -: none.

5.1. Ultraviolet (UV) Light

The antimicrobial properties of UV light are well investigated and, taking surface disinfection as an example, long established. However, the conventional application of UV light in the form of continuous exposure has numerous disadvantages. As the sanitizing effect is primarily attributed to the high ionizing energy of vacuum-UV (wavelength < 180 nm), the consumer acceptance is very low. The induced DNA damage, responsible for the microbial inhibition, also occurs in the sample which results in a substantial loss of quality. In addition, depending on the water content, substantial internal heating of the sample can occur. To avoid these unwanted side effects, recent research has focused on the application of pulsed UV light. In these studies, the treatment was carried out with numerous short flashes of light with a broad wavelength spectrum. Although the inhibitory effect was still attributed to the UV spectrum of light (wavelength 200–400 nm), microbial inhibition could be achieved with non-ionizing UV (>180 nm). At the same time, the undesired side effects were substantially reduced. Thus, pulsed UV light presents a novel, non-thermal, antimicrobial treatment with potential applications for food preservation. Therefore, this section is focused exclusively on the application of non-ionizing UV light. Microbial inactivation as a result of exposure to pulsed light, in vitro and in situ, has been reported by several studies and comprehensively reviewed by Oms-Oliu et al. [83].

Oms-Oliu et al. [83] demonstrated the successful inhibition of food spoilage fungi and bacteria with pulsed light when applied to various food matrices, including milk, honey, and fruits. However, only one available study investigated the application of pulsed light for pre-storage decontamination of cereals. Maftei et al. [76] reported up to a 4-log unit reduction for naturally occurring *Aspergillus* spp. in wheat grains. Treatment was carried out with 40 flashes of broad spectrum white light (180–1100 nm) with an overall energy release of 51.2 J/g. In addition, the authors reported that the same treatment with light of a narrower wavelength spectrum (305–1100 nm and 400–1100 nm, respectively) resulted in significantly less fungal inhibition.

Although pulsed UV light causes less damage to the sample than continuous UV irradiation, significantly reduced seed viability was reported nonetheless. Alongside the reduction of *Aspergillus* spp. from 10^5 to 10 cfu/g, the seed viability was significantly reduced by 15% [76]. Consequently, despite showing the potential for microbial decontamination, further optimisation of the processing conditions is required to improve efficiency and make the treatment potentially suitable for industrial application.

However, inhibition of common food spoilage bacteria was generally found to be more difficult, as reductions of no more than 1-log unit could be achieved without substantial impact on the sample quality [83]. Furthermore, no studies investigating the inhibition of bacteria commonly found on cereals or in cereal matrices are available. However, Nicorescu et al. [79] achieved up to a 1-log unit reduction of *Bacillus subtilis* in a liquid medium and artificially contaminated spices. After the treatment, approximately 10^4 cfu/g remained on the samples, despite an equivalent 90% reduction in the bacterial population. Thus, further research is needed to improve the antibacterial properties of pulsed UV light.

Several studies have also investigated the possibility of photodegradation of mycotoxins in vitro and in situ using UV and visible light. Treatment of different mycotoxins (OTA, OTB, and citrinin) in aqueous solution with light of various wavelengths (455, 470, 530, 590, and 627 nm) for five days resulted in significant degradation of all three toxins after exposure to the 455 nm and the 470 nm light [80]. In particular, light of wavelength 455 nm was found to be very efficient in terms of mycotoxin degradation. Subsequently, 455 nm light was applied in situ on artificially fungal-infected wheat kernels. After five days of exposure, the OTA and OTB levels were reduced by >90% compared to the untreated control [80].

In addition, the total aflatoxin content of various nuts could be substantially reduced by treatment with UV light (265 nm) for 15–45 min [81]. However, numerous factors were found to have a significant impact on the level of aflatoxin reduction. The most influential factors included the moisture content of the nuts, the toxin targeted, the exposure time, and the type of nut. The authors also reported higher resistance of AFB_1 and AFG_1 to the UV light compared to AFG_2. For all toxins, increasing sensitivity towards the treatment could be attributed to increasing moisture levels and exposure time.

To evaluate the potential use of photodegradation for the detoxification of cereals, the fate of the degraded mycotoxins also must be considered. Liu et al. [84] proposed a pathway for the UV light-induced photodegradation of AFB_1 after identification of the three main degradation products using UPLC-MS (ultra performance liquid chromatography-tandem mass spectrometer). Identification of the degradation products revealed that, from the two most important parts of the molecule in terms of toxicity, namely the terminal furan ring and the lactone ring (Table 1), only the latter was affected by photodegradation. In addition, the authors reported first order reaction kinetics and found the process to be independent of the initial toxin concentration (within 0.2–5 ppm) but directly related to the irradiation intensity. On the basis of these results, a subsequent study investigated the cytotoxicity of UV light degraded AFB_1 in an aqueous solution [85]. The authors assessed the toxicity of the three previously identified degradation products using the Ames test. Interestingly, the cytotoxicity was reduced by 40% compared to native AFB_1 but not fully eliminated. This leads to the question as to whether photodegradation can serve as a suitable tool for mycotoxin detoxification. Given that 60% of the initial cytotoxicity remained even after complete degradation of the toxin, alternative treatments are likely to be more suitable.

In conclusion, despite some potential for microbial decontamination and mycotoxin degradation, UV light appears to be difficult to apply without affecting sample quality. Although the use of pulsed light rather than continuous irradiation reduces the negative effects, no complete inhibition without quality loss has been reported. This applies in particular to the decontamination of food spoilage bacteria. For application on cereal grains, the uneven sample surface makes a reliable application even more challenging as a result of the shadow spots on the grains. In terms of mycotoxin degradation, it was shown that UV light was capable of total degradation of toxins such as AFB_1, but the degradation products were found to remain cytotoxic. Therefore, it appears an unsuitable method for mycotoxin detoxification.

5.2. Microwave Treatments

Microwave technology is widely used in the food industry and offers several advantages, including safety, high efficiency, and environmental protection, but often affects food quality [74]. Microwaves are defined as electromagnetic waves with frequencies ranging from 300 MHz to 300 GHz. The mechanism of microbial inhibition is primarily based on the internal heating of the sample resulting from molecular movement in the pulsing electromagnetic field. This leads to the denaturation of proteins, enzymes, and nucleic acids. This implies the risk of losing enzymatic activities in the grain, activities which are essential for downstream processing steps [74]. However, with optimised processing conditions, microwave treatment has the ability to fully inhibit microbial growth on cereal grains without compromising the grain's germination quality [86]. Therefore, the impact of various factors, such as moisture content, microbial load, or sample matrix, need to be investigated further to determine the optimal processing conditions for each cereal matrix.

Available literature suggests a microwave treatment for up to 10 min with an energy output of 1.45 kW with 2450 MHz results in a very minor reduction of total AFs, including AFB_1 [53]. However, the possibility of inhibiting the growth and spore germination of *Aspergillus* spp. in cereals and nuts by 3-log units, without significant quality deterioration, was reported by different authors [77,78]. A treatment time of 120 s with 2450 MHz and 1.25 kW was found to be sufficient. Consequently, due to the fungal inhibition, the amount of mycotoxins produced was also significantly lower.

Furthermore, non-thermal inactivation of microorganisms resulting from repeated exposure to sub-lethal doses of high frequency microwaves has been reported [74]. The lethal effect of low-dose microwave radiation (LDMR) on spoilage microorganisms was a result of a disruption of the cell membrane and induced DNA damage rather than protein denaturation. Thus, the mechanism by which LDMR causes fungal death is different from a conventional heating treatment [74]. However, in a study conducted on *A. parasiticus*, the severity of DNA injury was found to increase with rising temperatures. Thus, the inactivation effect is still partially related to the processing temperature. However, few available studies have investigated this topic. Thus, future research is needed to exploit this technology through the establishment of non-thermal, microwave-based, microbial hurdle processes that do not compromise the grain quality.

Overall, microwave treatment, because of internal heating, shows little potential for the microbial decontamination of cereals. This is primarily a result of heat-induced damage to the sample. The same conclusion must be drawn for the microwave-induced degradation of mycotoxins which possess high heat stability. However, the concept of non-thermal microbial inactivation appears much more promising. However, as research on this approach is in its infancy, it is not yet possible to predict the full potential and applicability to industrial grain decontamination process until more extensive research has been conducted.

5.3. Ultrasonication

The term 'ultrasound' (US) describes sonic waves with frequencies above the human audible range and are generally divided into two categories: high frequency ultrasound and power ultrasound. The former uses high frequencies of 2–20 MHz with low sound intensity (0.1–1 W/cm^2) and is predominantly used in food quality analysis and medical imaging. In contrast, power ultrasound, or high-intensity ultrasound, is characterised by low frequencies (20–100 kHz) but high sound intensity (10–1000 W/cm^2). Research on the inactivation of enzymes and microorganisms has predominantly focused on the application of high power ultrasound, considered a promising, novel, and non-thermal approach for microbial disinfection of various surfaces and food matrices [87]. In contrast, high frequency US shows much less potential for microbial decontamination. Therefore, this review focuses on power ultrasound exclusively.

Only a few studies have investigated the sole application of US e for microbial decontamination with contradictory results. Butz and Tauscher [88] demonstrated that US alone does not sufficiently inactivate food spoilage microorganisms, as the inactivating effects are not severe enough. In contrast, Chemat et al. [73] reported that, if the acoustic power applied is sufficiently high (>60 W/cm^2), even US alone can induce cell rupture and thus microbial inactivation. Successful decontamination using US was only achieved under laboratory conditions and would be difficult to apply on an industrial scale. Most data available suggests that US alone is a very inefficient and energy consuming treatment for microbial disinfection and therefore must be used in combination with other treatments, such as sanitizing chemicals or heat [89]. Furthermore, the generation of such high-power US requires an immense energy input and is relatively inefficient compared to other techniques. This is further supported by O'Donnell et al. [19] who described the challenges encountered by the industrial scale-up of US technology. In addition, the application of this technology to cereal grains could prove difficult, as the treatment has to be carried out in a liquid medium. Therefore, it would be crucial that the cavitation of the liquid around the grains is evenly distributed.

In general, the biggest potential of US is in combination with mild heat or sanitizing agents, which have been shown to have a synergistic effect by several authors [90,91]. Herceg et al. [75] achieved total inhibition of *Aspergillus* and *Penicillium* spp., after exposure to ultrasound for at least 6 min at 60 °C when the applied power was ranging between 20 and 39 W/cm^2. However, these results were achieved in liquid sample matrices in which the US waves can easily travel, resulting in an evenly distributed cavitation effect. But for possible application on cereal crops, it remains unclear if enough cavitation throughout the whole sample can be generated for a noteworthy disinfection effect. Thus, a small sample size appears necessary to provide uniform cavitation. Furthermore, the grains would be required to be in a "washing solution", which produces a further challenge to the industrial use of this technology.

No available studies have investigated the application of US to decontaminate solid foods. Chemat et al. [73] recently reviewed the application of US in the food industry and its advances for decontamination of liquid food systems. However, chemical disinfection supported by US is likely to result in synergistic effects. Thus, it is apparently a promising improvement over conventional chemical disinfection (reduction of treatment time and unpleasant side effects) and therefore should be considered for further investigations. Ultrasound could be used to provide the energy necessary to form free radicals as reactive species and so support the disinfection efficiency of commonly used surface sanitizers. In particular, hydrogen peroxide- or sodium hypochlorite-based disinfection of food or food contact materials can be efficiently supported by US, creating more hydroxyl- and hypochlorite radicals, respectively. Furthermore, the use of US can substantially increase the speed and efficacy of conventional food preservation methods such as sterilisation and pasteurisation. This would allow a reduction in the processing time and temperature and therefore reduce undesirable side effects, such as changes in taste, colour, and nutritional value [73].

Few studies have investigated the impact of ultrasonication on mycotoxins. Lindner and Hasenhuti [82] reported the successful degradation of trichothecenes in contaminated corn while the technological performance of the treated samples remained unchanged. However, as discussed earlier, for the combination of gamma irradiation with chemical sanitizers, the production of hydroxyl radicals can lead to chemical degradation of aflatoxins. Theoretically it should be possible to produce such radicals using US. However, to the best of the authors' knowledge, no studies have been conducted on this topic.

On the basis of published research, clearly only a combination of US with heat or chemical sanitizers can sufficiently inactivate vegetative cells, spores, and enzymes simultaneously. Only then can consistent and high product quality be ensured, as the microbial enzymes can cause great damage to proteins and carbohydrates of the grains, even after killing of the vegetative cells [7].

6. High Hydrostatic Pressure (HHP)

Another emergent approach for the decontamination of food and feed products from spoilage microbes is treatment with high hydrostatic pressure (HHP). Well-established applications include the preservation of meat products, oysters, fruit juices, and many ready-to-eat foods. HHP may inactivate vegetative microorganisms and fungal spores at relatively low temperatures without compromising sensory and nutritional properties [92,93]. Thus, it has potential use in the expansion of the production of value-added foods.

Microbial inactivation is reported for processing pressures ranging from 100 to 800 MPa and relatively short times (a few seconds to several minutes). Combined treatment with mild temperatures between 20 and 50 °C to inactivate enzymes also have been reported. The treatment conditions depend fundamentally on the food matrix as well as the microorganisms and enzymes targeted. However, it is noteworthy that this technology cannot inactivate bacterial endospores with the application of the processing parameters commonly used in the food industry [94]. Pressures of at least 600 MPa with mild temperatures (60 °C) are required for the killing of spores of most food spoilage bacteria. Certain strains of *Clostridium botulinum* and *Bacillus species* are reported to withstand hydrostatic pressures of up to 1000 MPa [92]. Therefore, they present a much bigger challenge and cannot be inhibited by HHP alone, but which may be possible in combination with more severe heat treatment [95].

For the inactivation of *Escherichia coli* K-12 and *Staphylococcus aureus* ATCC 6538 in cheese slurries, 20 min of 400 MPa at 30 °C and of 600 MPa at 20 °C, respectively, were found to be sufficient [96]. Likewise, potential pathogenic food bacteria were inhibited due to HHP treatment. Studies on almonds, pressurised in water for 6 min at 414 MPa, followed by air drying at room temperature or 115 °C for 25 min, resulted in bacterial growth reduction of 4- and 6.7-log units, respectively [97]. The authors concluded that HHP treatments show great potential for microbial inactivation if the sample is suspended in the pressurizing medium.

With regards to fungal inhibition by HHP, Martínez-Rodríguez et al. [98] investigated the impact of HHP (300, 400, and 500 MPa, respectively) at 20 °C for 10 min on fungal mycelia development, spore viability, and enzymatic activity of *P. roqueforti*. Mycelia development was significantly reduced following all three treatments but in direct correlation with the applied pressure. Furthermore, the spore viability was notably reduced after exposure to 300 MPa and completely inhibited at higher pressures. Similarly, the total lipolytic activity of the samples decreased with increasing pressure. Researchers reported similar results for the inhibition of different food spoilage fungi, such as *Penicillium* spp., *Fusarium* spp., and *Aspergillus* spp., in a liquid medium and cheese [96,99]. This suggests that HHP treatments are a promising option for the decontamination of cereal grains in particular, as the grains are known to withstand pressures of 400 MPa without sensory or grain quality deterioration. Thus, potential spoilage fungi and their hydrolytic enzymes could be completely inhibited, without the need for classic heat treatments or chemicals. However, asco-spores of heat-resistant moulds were also reported to possess a high resistance to HHP. Hence, pressure treatments routinely applied to foods do not result in sufficient inhibition. However, HHP appears to sensitize asco-spores to subsequent heat treatments. Thus, a combination of heat and pressure treatment appears very promising [99].

Although the prevention of fungal growth is the best way to ensure mycotoxin-free crops, minor in-field contamination can result in mycotoxins present in otherwise good quality cereals. Thus, the in vitro and in situ degradation of mycotoxins is another relevant topic of research. However, few studies have investigated the sensitivity of mycotoxins to HHP treatments. The degradation of patulin in fruit juices was reported after treatment with 600 MPa for 300 s at 11 °C [100]. Unfortunately, there are no studies available regarding HHP treatment of cereals for microbial decontamination or degradation of mycotoxins. In addition, as observed previously by several researchers, the sensitivity of pathogenic and spoilage organisms and their metabolic products greatly depends on a number of factors, including the surrounding sample matrix, microbial strain, processing conditions, and moisture levels [92,95,96,99,101,102]. Therefore, without sufficient studies on cereal grains and investigations of typical cereal contaminants, it is impossible to predict the efficiency of HHP treatments on the microbial decontamination of cereal grains prior to storage.

In conclusion, the application of HHP appears to represent a promising approach for ensuring microbial safety without the need for chemical preservatives. However, HHP treatments require the sample to be suspended in a liquid medium. Otherwise, the pressure applied cannot be distributed evenly, leading to unsatisfactory results. This ultimately would require high moisture levels in the stored grains or would necessitate an additional drying step after the initial HHP treatment. HHP treatment has been shown to be effective in reducing the microbial load of foods for both pathogenic and spoilage microorganisms with minimal impact on the product quality. However, that the treatment is unsuitable for in-line procedures and must be performed in a batch process presents its biggest draw-back. Various parameters such as pressure, time, temperature, and pH have to be considered to optimise the process in terms of both microbial inactivation and in consideration of final product quality [103].

In particular, the combination of HHP with heating treatments requires further research to fully exploit its potential in the development of new products.

7. Conclusion and Future Trends

Scientific evidence for the potential of each treatment as a tool for microbial decontamination is available. In particular, novel technologies not currently used in industry were found to present several advantages over established ones. For example, the use of superheated steam combines a faster, more efficient microbial decontamination with less impact on the grain quality compared to conventional saturated steam or hot air treatments. In addition, the use of e-beam irradiation, high hydrostatic pressure, and microwaves, based on non-thermal inactivation mechanisms, presents several advantages over more established technologies, such as gamma rays or UV light. However, due to their novelty, particularly in terms of microbial decontamination of cereals, the side effects of these technologies have been sparsely investigated. Therefore, further research is required to better understand the impact on the treated grains and the effect on spoilage organisms after exposure to sub-lethal doses.

Another topic investigated in this review was the degradation of in-field produced and accumulated mycotoxins to prevent a potential consumer health hazard. All the evaluated treatments showed some potential for reducing the mycotoxin content in cereals. However, as a result of their structural diversity (Table 1) and higher intrinsic resistance against many treatments, when compared to the living organisms, the reduction of the myxotoxigenic load was found to be a major challenge. In many cases, the treatment time and intensity had to be much higher for mycotoxin degradation than for the inhibition of living organisms. Thus, successful degradation of these toxins is a major challenge without compromising the grain's quality and seed viability. In addition, major concern related to the decay of mycotoxins are the degradation products released. The vast majority of research has examined the reduction in the level of the original parent mycotoxin, paying little attention to the degradation products or mechanism of degradation. Therefore, the identity of the degradation products and their toxicity is unknown. As a result, the efficiency of mycotoxin degradation cannot be evaluated as the released compounds may potentially be equal or even more toxic compared to the parent toxin. Future research will need to understand the possible degradation mechanisms to identify the resulting compounds. Only then will it be possible to assess their toxicity and evaluate the success and efficiency of the treatments discussed in this review. Furthermore, concerns regarding a possible reformation of the toxins during subsequent processing steps and the formation of masked mycotoxins due to the treatments merits greater research interest.

For both purposes, microbial decontamination and detoxification, each treatment encompasses associated difficulties for the successful application to cereals. The even and homogeneous treatment of the whole sample appears to be the biggest challenge. In addition, the efficiency of a treatment depends on various influential factors. The sample matrix, target organism, moisture content, and water availability were the most frequently observed influencing factors within the available literature. Thus, the treatment conditions and setting must be optimised for each crop to make industrial application possible. However, even under ideal process conditions it appears unlikely that one physical treatment can result in sufficient microbial decontamination and detoxification without substantial grain quality deterioration. Consequently, the combination of several treatments appears to represent the most promising approach for optimal results. Future research should therefore focus on understanding and the optimisation of the synergies which are likely to be achievable through combinatory treatments. Only then will it be possible to produce products that meet the highest standards in terms of food quality and safety.

Acknowledgments: Financial support for this research was awarded by the Irish Government under the National Development Plan 2007–2013 through the research program FIRM/RSF/CoFoRD. This research was also partly funded by the Irish Department of Agriculture, Food and the Marine.

Author Contributions: M.S., E.Z. and E.K.A. conveived and designed the work; M.S. wrote the review; E.Z. and E.K.A. proof read and revised the article.

Conflicts of Interest: The authors declare no conflict of interest.

References

1. Freita-Silva, O.; de Oliveira, P.S.; Freire Júnior, M. Potential of Electron Beams to Control Mycotoxigenic Fungi in Food. *Food Eng. Rev.* **2014**, 160–170. [CrossRef]
2. Oerke, E.-C. Crop losses to pests. *J. Agric. Sci.* **2006**, *144*, 31. [CrossRef]
3. Siciliano, I.; Spadaro, D.; Prelle, A.; Vallauri, D.; Cavallero, M.C.; Garibaldi, A.; Gullino, M.L. Use of cold atmospheric plasma to detoxify hazelnuts from aflatoxins. *Toxins* **2016**, *8*, 125. [CrossRef] [PubMed]
4. Magan, N.; Hope, R.; Cairns, V.; Aldred, D. Post-harvest fungal ecology: Impact of fungal growth and mycotoxin accumulation in stored grain. *Eur. J. Plant Pathol.* **2003**, *109*, 723–730. [CrossRef]
5. Oliveira, P.M.; Zannini, E.; Arendt, E.K. Cereal fungal infection, mycotoxins, and lactic acid bacteria mediated bioprotection: From crop farming to cereal products. *Food Microbiol.* **2014**, *37*, 78–95. [CrossRef] [PubMed]

6. Magan, N.; Aldred, D. Post-Harvest Control Strategies: Minimizingmycotoxins in the Food Chain. *Int. J. Food Microbiol.* **2007**, *119*, 131–139. [CrossRef] [PubMed]

7. Schmidt, M.; Horstmann, S.; De Colli, L.; Danaher, M.; Speer, K.; Zannini, E.; Arendt, E.K. Impact of fungal contamination of wheat on grain quality criteria. *J. Cereal Sci.* **2016**, *69*, 95–103. [CrossRef]

8. Audenaert, K.; Monbaliu, S.; Deschuyffeleer, N.; Maene, P.; Vekeman, F.; Haesaert, G.; De Saeger, S.; Eeckhout, M. Neutralized electrolyzed water efficiently reduces *Fusarium* spp. in vitro and on wheat kernels but can trigger deoxynivalenol (DON) biosynthesis. *Food Control* **2012**, *23*, 515–521. [CrossRef]

9. Cheli, F.; Pinotti, L.; Rossi, L.; Dell'Orto, V. Effect of milling procedures on mycotoxin distribution in wheat fractions: A review. *LWT—Food Sci. Technol.* **2013**, *54*, 307–314. [CrossRef]

10. Vidal, A.; Marín, S.; Ramos, A.J.; Cano-Sancho, G.; Sanchis, V. Determination of aflatoxins, deoxynivalenol, ochratoxin A and zearalenone in wheat and oat based bran supplements sold in the Spanish market. *Food Chem. Toxicol.* **2013**, *53*, 133–138. [CrossRef] [PubMed]

11. Magan, N.; Aldred, D. *Food Spoilage Microorganisms*; Woodhead Publishing: Sawston, UK, 2006.

12. Balasubramaniam (Bala), V.M.; Martínez-Monteagudo, S.I.; Gupta, R. Principles and Application of High Pressure–Based Technologies in the Food Industry. *Annu. Rev. Food Sci. Technol.* **2015**, *6*, 435–462. [CrossRef] [PubMed]

13. Crowley, S.; Mahony, J.; Van Sinderen, D. Current perspectives on antifungal lactic acid bacteria as natural bio-preservatives. *Trends Food Sci. Technol.* **2013**, *33*, 93–109. [CrossRef]

14. Pawlowska, A.M.; Zannini, E.; Coffey, A.; Arendt, E.K. "Green Preservatives": Combating Fungi in the Food and Feed Industry by Applying Antifungal Lactic Acid Bacteria. *Adv. Food Nutr. Res.* **2012**, *66*, 217–238. [CrossRef] [PubMed]

15. Magan, N.; Lacey, J. Effects of gas composition and water activity on growth of field and storage fungi and their interactions. *Trans. Br. Mycol. Soc.* **1984**, *82*, 305–314. [CrossRef]

16. Taniwaki, M.H.; Hocking, A.D.; Pitt, J.I.; Fleet, G.H. Growth of fungi and mycotoxin production on cheese under modified atmospheres. *Int. J. Food Microbiol.* **2001**, *68*, 125–133. [CrossRef]

17. Gupta, A.; Sinha, S.N.; Atwal, S.S. Modified Atmosphere Technology in Seed Health Management: Laboratory and Field Assay of Carbon Dioxide against Storage Fungi in Paddy. *Plant Pathol. J.* **2014**, *13*, 193–199. [CrossRef]

18. Savi, G.D.; Scussel, V.M. Effects of Ozone Gas Exposure on Toxigenic Fungi Species from *Fusarium*, *Aspergillus*, and *Penicillium* Genera. *Ozone Sci. Eng. J. Int. Ozone Assoc.* **2014**, *36*, 144–152. [CrossRef]

19. O'Donnell, C.P.; Tiwari, B.K.; Bourke, P.; Cullen, P.J. Effect of ultrasonic processing on food enzymes of industrial importance. *Trends Food Sci. Technol.* **2010**, *21*, 358–367. [CrossRef]

20. Klein, J.D.; Lurie, S. Postharvest heat treatment and fruit quality. *Postharvest News Inf.* **1991**, *2*, 15–19.

21. Lan, S. Effects of Post-Harvest Treatment and Heat Stress on the Antioxidant Properties of Wheat. Master's Thesis, University of Maryland, College Park, MD, USA, 3 August 2006.

22. Lehtinen, P.; Kiiliäinen, K.; Lehtomäki, I.; Laakso, S. Effect of Heat Treatment on Lipid Stability in Processed Oats. *J. Cereal Sci.* **2003**, *37*, 215–221. [CrossRef]

23. Rose, D.J.; Bianchini, A.; Martinez, B.; Flores, R.A. Methods for reducing microbial contamination of wheat flour and effects on functionality. *Cereal Foods World* **2012**, *57*, 104–109. [CrossRef]

24. Bari, L.; Ohki, H.; Nagakura, K.; Ukai, M. Application of Ultra Superheated Steam Technology (USST) to Food Grain Preservation at Ambient Temperature for Extended Periods of Time. *Adv. Food Technol. Nutr. Sci.* **2015**, *SE1*, S14–S21. [CrossRef]

25. Gilbert, J.; Woods, S.M.; Turkington, T.K.; Tekauz, A. Effect of heat treatment to control *Fusarium graminearum* in wheat seed. *Can. J. Plant Pathol.* **2005**, *27*, 448–452. [CrossRef]

26. Chang, Y.; Li, X.-P.; Liu, L.; Ma, Z.; Hu, X.; Zhao, W.; Gao, G. Effect of Processing in Superheated Steam on Surface Microbes and Enzyme Activity of Naked Oats. *J. Food Process. Preserv.* **2015**, *39*, 2753–2761. [CrossRef]

27. Hu, Y.; Nie, W.; Hu, X.; Li, Z. Microbial decontamination of wheat grain with superheated steam. *Food Control* **2016**, *62*, 264–269. [CrossRef]

28. Ban, G.-H.; Kang, D.-H. Effectiveness of superheated steam for inactivation of *Escherichia coli* O157:H7, *Salmonella* Typhimurium, *Salmonella* Enteritidis phage type 30, and *Listeria monocytogenes* on almonds and pistachios. *Int. J. Food Microbiol.* **2016**, *220*, 19–25. [CrossRef] [PubMed]

29. Cenkowski, S.; Pronyk, C.; Zmidzinska, D.; Muir, W.E. Decontamination of food products with superheated steam. *J. Food Eng.* **2007**, *83*, 68–75. [CrossRef]

30. Clear, R.M.; Patrick, S.K.; Wallis, R.; Turkington, T.K. Effect of dry heat treatment on seed-borne *Fusarium graminearum* and other cereal pathogens. *Can. J. Plant Pathol.* **2002**, *24*, 489–498. [CrossRef]

31. Bond, W.W.; Favero, M.S.; Petersen, N.J.; Marshall, J.H. Dry-heat inactivation kinetics of naturally occurring spore populations. *Appl. Microbiol.* **1970**, *20*, 573–578. [PubMed]

32. Nielsen, K.F.; Holm, G.; Uttrup, L.P.; Nielsen, P.A. Mould growth on building materials under low water activities. Influence of humidity and temperature on fungal growth and secondary metabolism. *Int. Biodeterior. Biodegrad.* **2004**, *54*, 325–336. [CrossRef]

33. Miller, J.D.; Trenholm, H.L. *Mycotoxins in Grain Compounds Other than Aflatoxin*; Eagan Press: Saint Paul, MN, USA, 1997.

34. Bretz, M.; Knecht, A.; Göckler, S.; Humpf, H.U. Structural elucidation and analysis of thermal degradation products of the *Fusarium* mycotoxin nivalenol. *Mol. Nutr. Food Res.* **2005**, *49*, 309–316. [CrossRef] [PubMed]

35. Vidal, A.; Sanchis, V.; Ramos, A.J.; Marín, S. Thermal stability and kinetics of degradation of deoxynivalenol, deoxynivalenol conjugates and ochratoxin A during baking of wheat bakery products. *Food Chem.* **2015**, *178*, 276–286. [CrossRef] [PubMed]

36. Boudra, H.; Lebars, P.; Lebars, J. Thermostability of Ochratoxin A in Wheat under 2 Moisture Conditions. *Appl. Environ. Microbiol.* **1995**, *61*, 1156–1158. [PubMed]

37. Syamaladevi, R.M.; Tang, J.; Villa-Rojas, R.; Sablani, S.; Carter, B.; Campbell, G. Influence of Water Activity on Thermal Resistance of Microorganisms in Low-Moisture Foods: A Review. *Compr. Rev. Food Sci. Food Saf.* **2016**, *15*, 353–370. [CrossRef]

38. Ban, G.H.; Yoon, H.; Kang, D.H. A comparison of saturated steam and superheated steam for inactivation of *Escherichia coli* O157: H7, *Salmonella* Typhimurium, and *Listeria monocytogenes* biofilms on polyvinyl chloride and stainless steel. *Food Control* **2014**, *40*, 344–350. [CrossRef]

39. Alfy, A.; Kiran, B.V.; Jeevitha, G.C.; Hebbar, H.U. Recent Developments in Superheated Steam Processing of Foods—A Review. *Crit. Rev. Food Sci. Nutr.* **2016**, *56*, 2191–2208. [CrossRef] [PubMed]

40. Jalili, M. A Review on Aflatoxins Reduction in Food. *Iran. J. Health Saf. Environ.* **2015**, *3*, 445–459.

41. Nemţanu, M.R.; Braşoveanu, M.; Karaca, G.; Erper, I. Inactivation effect of electron beam irradiation on fungal load of naturally contaminated maize seeds. *J. Sci. Food Agric.* **2014**, *94*, 2668–2673. [CrossRef] [PubMed]

42. Dev, S.R.S.; Birla, S.L.; Raghavan, G.S.V.; Subbiah, J. Microbial decontamination of food by microwave (MW) and radiao frequency (RF). In *Microbial Decontamination in the Food Industry*; Novel Methods and Applications; Woodhead Publishing: Sawston, UK, 2012; pp. 274–299.

43. FAO/IAEA Training Manual on Food Irradiation Technology and Techniques. In *Technical Reports Series, Proceedings of the International Atomic Energy Commission*; IAEA Publications: Vienna, Austria, 1982; p. 224.

44. Mahmoud, N.S.; Awad, S.H.; Madani, R.M.A.; Osman, F.A.; Elmamoun, K.; Hassan, A.B. Effect of γ radiation processing on fungal growth and quality characteristcs of millet grains. *Food Sci. Nutr.* **2016**, *4*, 342–347. [CrossRef] [PubMed]

45. D'Ovidio, K.L.; Trucksess, M.W.; Devries, J.W.; Bean, G. Effects of irradiation on fungi and fumonisin B1 in corn, and of microwave-popping on fumonisins in popcorn. *Food Addit. Contam.* **2007**, *24*, 735–743. [CrossRef] [PubMed]

46. Lung, H.-M.; Cheng, Y.-C.; Chang, Y.-H.; Huang, H.-W.; Yang, B.B.; Wang, C.-Y. Microbial decontamination of food by electron beam irradiation. *Trends Food Sci. Technol.* **2015**, *44*, 66–78. [CrossRef]

47. Supriya, P.; Sridhar, K.R.; Ganesh, S. Fungal decontamination and enhancement of shelf life of edible split beans of wild legume Canavalia maritima by the electron beam irradiation. *Radiat. Phys. Chem.* **2014**, *96*, 5–11. [CrossRef]

48. Salem, E.A.; Soliman, S.A.; El-Karamany, A.M.; El-shafea, Y.M.A. Effect of Utilization of Gamma Radiation Treatment and Storage on Total Fungal Count, Chemical Composition and Technological Properties Wheat Grain. *Egypt. J. Biol. Pest Control* **2016**, *26*, 163–171.

49. Aziz, N.H.; El-Far, F.M.; Shahin, A.A.M.; Roushy, S.M. Control of *Fusarium* moulds and fumonisin B1 in seeds by gamma-irradiation. *Food Control* **2007**, *18*, 1337–1342. [CrossRef]

50. Carocho, M.; Antonio, A.L.; Barreira, J.C.M.; Rafalski, A.; Bento, A.; Ferreira, I.C.F.R. Validation of Gamma and Electron Beam Irradiation as Alternative Conservation Technology for European Chestnuts. *Food Bioprocess Technol.* **2014**, *7*, 1917–1927. [CrossRef]

51. Carocho, M.; Barros, L.; Antonio, A.L.; Barreira, J.C.M.; Bento, A.; Kaluska, I.; Ferreira, I.C.F.R. Analysis of organic acids in electron beam irradiated chestnuts (*Castanea sativa* Mill.): Effects of radiation dose and storage time. *Food Chem. Toxicol.* **2013**, *55*, 348–352. [CrossRef] [PubMed]

52. Akueche, E.C.; Anjorin, S.T.; Harcourt, B.I.; Kana, D. Studies on fungal load, total aflatoxins and ochratoxin a contents of gamma-irradiated and non-irradiated Sesamum indicum grains from Abuja markets, Nigeria. *Kasetsart J. Nat. Sci.* **2012**, *46*, 371–382.

53. Herzallah, S.; Alshawabkeh, K.; Al Fataftah, A. Aflatoxin decontamination of artificially contaminated feeds by sunlight, γ-radiation, and microwave heating. *J. Appl. Poult. Res.* **2008**, *17*, 515–521. [CrossRef]

54. Jalili, M.; Jinap, S.; Noranizan, M.A. Aflatoxins and ochratoxin a reduction in black and white pepper by gamma radiation. *Radiat. Phys. Chem.* **2012**, *81*, 1786–1788. [CrossRef]

55. Mehrez, A.; Maatouk, I.; Romero-González, R.; Ben Amara, A.; Kraiem, M.; Garrido Frenich, A.; Landoulsi, A. Assessment of ochratoxin A stability following gamma irradiation: Experimental approaches for feed detoxification perspectives. *World Mycotoxin J.* **2016**, *9*, 289–298. [CrossRef]

56. Hooshmand, H.; Klopfenstein, C.F. Effects of gamma irradiation on mycotoxin disappearance and amino acid contents of corn, wheat, and soybeans with different moisture contents. *Plant Foods Hum. Nutr.* **1995**, *47*, 227–238. [CrossRef] [PubMed]

57. Carocho, M.; Barreira, J.C.; Antonio, A.L.; Bento, A.; Kaluska, I.; Ferreira, I.C. Effects of electron beam radiation on nutritional parameters of Portuguese chestnuts (*Castanea sativa* mill.). *J. Agric. Food Chem.* **2012**, *60*, 7754–7760. [CrossRef] [PubMed]

58. El-Naggar, S.M.; Mikhaiel, A.A. Disinfestation of stored wheat grain and flour using gamma rays and microwave heating. *J. Stored Prod. Res.* **2011**, *47*, 191–196. [CrossRef]

59. Melki, M.; Marouani, A. Effects of gamma rays irradiation on seed germination and growth of hard wheat. *Environ. Chem. Lett.* **2010**, *8*, 307–310. [CrossRef]

60. Deberghes, P.; Betbeder, A.M.; Boisard, F.; Blanc, R.; Delaby, J.F.; Krivobok, S.; Steiman, R.; Seigle-Murandi, F.; Creppy, E.E. Detoxification of ochratoxin A, a food contaminant: Prevention of growth of *Aspergillus ochraceus* and its production of ochratoxin A. *Mycotoxin Res.* **1995**, *11*, 37–47. [CrossRef] [PubMed]

61. Di Stefano, V.; Pitonzo, R.; Cicero, N.; D'Oca, M.C. Mycotoxin contamination of animal feedingstuff: Detoxification by gamma-irradiation and reduction of aflatoxins and ochratoxin A concentrations. *Food Addit. Contam. Part A* **2014**, *31*, 2034–2039. [CrossRef] [PubMed]

62. Aquino, S.; Ferreira, F.; Ribeiro, D.H.B.; Corrêa, B.; Greiner, R.; Villavicencio, A.L.C.H. Evaluation of viability of *Aspergillus flavus* and aflatoxins degradation in irradiated samples of maize. *Braz. J. Microbiol.* **2005**, *36*, 352–356. [CrossRef]

63. Farag, R.S.; El-Baroty, G.S.; Abo-Hagger, A.A. Aflatoxin destruction and residual toxicity of contaminated-irradiated yellow corn and peanuts on rats. *Adv. Food Sci.* **2004**, *26*, 122–129.

64. Prado, G.; de Carvalho, E.P.; Oliveira, M.S.; Madeira, J.G.C.; Morais, V.D.; Correa, R.F.; Cardoso, V.N.; Soares, T.V.; da Silva, J.F.M.; Gonçalves, R.C.P. Effect of gamma irradiation on the inactivation of aflatoxin B1 and fungal flora in peanut. *Braz. J. Microbiol.* **2003**, *34*, 138–140. [CrossRef]

65. Wang, F.; Xie, F.; Xue, X.; Wang, Z.; Fan, B.; Ha, Y. Structure elucidation and toxicity analyses of the radiolytic products of aflatoxin B_1 in methanol-water solution. *J. Hazard. Mater.* **2011**, *192*, 1192–1202. [CrossRef] [PubMed]

66. Kume, T.; Furuta, M.; Todoriki, S.; Uenoyama, N.; Kobayashi, Y. Status of food irradiation in the world. *Radiat. Phys. Chem.* **2009**, *78*, 222–226. [CrossRef]

67. Kottapalli, B.; Wolf-Hall, C.E.; Schwarz, P. Effect of electron-beam irradiation on the safety and quality of *Fusarium*-infected malting barley. *Int. J. Food Microbiol.* **2006**, *110*, 224–231. [CrossRef] [PubMed]

68. Stepanik, T.; Kost, D.; Nowicki, T.; Gaba, D. Effects of electron beam irradiation on deoxynivalenol levels in distillers dried grain and solubles and in production intermediates. *Food Addit. Contam.* **2007**, *24*, 1001–1006. [CrossRef] [PubMed]

69. Lanza, C.M.; Mazzaglia, A.; Paladino, R.; Auditore, L.; Barnà, D.; Loria, D.; Trifirò, A.; Trimarchi, M.; Bellia, G. Characterization of peeled and unpeeled almond (*Prunus amygdalus*) flour after electron beam processing. *Radiat. Phys. Chem.* **2013**, *86*, 140–144. [CrossRef]

70. Schoeller, N.P.; Ingham, S.C.; Ingham, B.H. Assessment of the Potential for *Listeria monocytogenes* Survival and Growth during Alfalfa Sprout Production and Use of Ionizing Radiation as a Potential Intervention Treatment. *J. Food Prot.* **2002**, *8*, 1259–1266. [CrossRef]

71. Kikuchi, O.K.; Todoriki, S.; Saito, M.; Hayashi, T. Efficacy of soft-electron (low-energy electron beam) for soybean decontamination in comparison with gamma-rays. *J. Food Sci.* **2003**, *68*, 649–652. [CrossRef]

72. Farkas, J.; Mohacsi-Farkas, C. History and future of food irradiation. *Trends Food Sci. Technol.* **2011**, *22*, 121–126. [CrossRef]

73. Oms-Oliu, G.; Martín-Belloso, O.; Soliva-Fortuny, R. Pulsed light treatments for food preservation. A review. *Food Bioprocess Technol.* **2010**, *3*, 13–23. [CrossRef]

74. Aron Maftei, N.; Ramos-Villarroel, A.Y.; Nicolau, A.I.; Martín-Belloso, O.; Soliva-Fortuny, R. Pulsed light inactivation of naturally occurring moulds on wheat grain. *J. Sci. Food Agric.* **2014**, *94*, 721–726. [CrossRef] [PubMed]

75. Nicorescu, I.; Nguyen, B.; Moreau-Ferret, M.; Agoulon, A.; Chevalier, S.; Orange, N. Pulsed light inactivation of Bacillus subtilis vegetative cells in suspensions and spices. *Food Control* **2013**, *31*, 151–157. [CrossRef]

76. Schmidt-Heydt, M.; Cramer, B.; Graf, I.; Lerch, S.; Hunpf, H.-U.; Geisen, R. Wavelength-dependent degradation of ochratoxin and citrinin by light in vitro and in vivo and its implications on *Penicillium*. *Toxins* **2012**, *4*, 1535–1551. [CrossRef] [PubMed]

77. Jubeen, F.; Bhatti, I.A.; Khan, M.Z.; Shahid, M. Effect of UVC Irradiation on Aflatoxins in Ground Nut (*Arachis hypogea*) and Tree Nuts (*Juglans regia, Prunus duclus* and *Pistachio vera*). *Chem. Soc. Pak.* **2012**, *34*, 1–10.

78. Liu, R.; Jin, Q.; Tao, G.; Shan, L.; Huang, J.; Liu, Y.; Wang, X.; Mao, W.; Wang, S. Photodegradation kinetics and byproducts identification of the Aflatoxin B1 in aqueous medium by ultra-performance liquid chromatography-quadrupole time-of-flight mass spectrometry. *J. Mass Spectrom.* **2010**, *45*, 553–559. [CrossRef] [PubMed]

79. Liu, R.; Chang, M.; Jin, Q.; Huang, J.; Liu, Y.; Wang, X. Degradation of aflatoxin B1 in aqueous medium through UV irradiation. *Eur. Food Res. Technol.* **2011**, *233*, 1007–1012. [CrossRef]

80. Fang, Y.; Hu, J.; Xiong, S.; Zhao, S. Effect of low-dose microwave radiation on *Aspergillus parasiticus*. *Food Control* **2011**, *22*, 1078–1084. [CrossRef]

81. Ursu, M.-P. Usage of Microwaves for Decontamination of Sensible Materials and Cereal Seeds. *Rev. Tehnol. Neconv.* **2015**, *19*, 60–64.

82. Kabak, B.; Dobson, A.D.; Var, I. Strategies to Prevent Mycotoxin Contamination of Food and Animal Feed: A Review. *Crit. Rev. Food Sci. Nutr.* **2006**, *46*, 593–619. [CrossRef] [PubMed]

83. Basaran, P.; Akhan, Ü. Microwave irradiation of hazelnuts for the control of aflatoxin producing *Aspergillus parasiticus*. *Innov. Food Sci. Emerg. Technol.* **2010**, *11*, 113–117. [CrossRef]

84. Feng, H.; Yang, W.; Hielscher, T. Power Ultrasound. *Food Sci. Technol. Int.* **2008**, *14*, 433–436. [CrossRef]

85. Butz, P.; Tauscher, B. Emerging technologies: Chemical aspects. *Food Res. Int.* **2002**, *35*, 279–284. [CrossRef]

86. Chemat, F.; Zill-E-Huma; Khan, M.K. Applications of ultrasound in food technology: Processing, preservation and extraction. *Ultrason. Sonochem.* **2011**, *18*, 813–835. [CrossRef] [PubMed]

87. Bilek, S.E.; Turantaş, F. Decontamination efficiency of high power ultrasound in the fruit and vegetable industry, a review. *Int. J. Food Microbiol.* **2013**, *166*, 155–162. [CrossRef] [PubMed]

88. Scouten, A.J.; Beuchat, L.R. Combined effects of chemical, heat and ultrasound treatments to kill *Salmonella* and *Escherichia coli* O157:H7 on alfalfa seeds. *J. Appl. Microbiol.* **2002**, *92*, 668–674. [CrossRef] [PubMed]

89. Seymour, I.J.; Burfoot, D.; Smith, R.L.; Cox, L.A.; Lockwood, A. Ultrasound decontamination of minimally processed fruits and vegetables. *Int. J. Food Sci. Technol.* **2002**, *37*, 547–557. [CrossRef]

90. Herceg, Z.; Jambrak, R.R.; Vukušić, T.; Stulić, V.; Stanzer, D.; Milošević, S. The effect of high-power ultrasound and gas phase plasma treatment on *Aspergillus* spp. and *Penicillium* spp. count in pure culture. *J. Appl. Microbiol.* **2015**, *118*, 132–141. [CrossRef] [PubMed]

91. Lindner, W.; Hasenhuti, K. Decontamination and Detoxification of Corn Which Was Contaminated with Trichothecenes Applying Ultrasonication (Abstr.). In Proceedings of the IX Internat IUPAC Symposium on Mycotoxins and Phytotoxins, Rome, Italy, 7–31 May 1996; p. 182.

92. Heinz, V.; Buckow, R. Food preservation by high pressure. *J. Verbrauch. Lebensmittelsich.* **2009**, *5*, 73–81. [CrossRef]

93. Polydera, A.C.; Stoforos, N.G.; Taoukis, P.S. Comparative shelf life study and vitamin C loss kinetics in pasteurised and high pressure processed reconstituted orange juice. *J. Food Eng.* **2003**, *60*, 21–29. [CrossRef]

94. Wannasawat Ratphitagsanti, M. *Approaches for Enhancing Lethality of Bacterial Spores Treated by Pressure-Assisted Thermal Processing*; ProQuest Dissertations Publishing: Ann Arbor, MI, USA, 2009.

95. Patterson, M.F. Microbiology of pressure-treated foods. *J. Appl. Microbiol.* **2005**, *98*, 1400–1409. [CrossRef] [PubMed]

96. O'Reilly, C.E.; O'Connor, P.M.; Kelly, A.L.; Beresford, T.P.; Murphy, P.M. Use of hydrostatic pressure for inactivation of microbial contaminants in cheese. *Appl. Environ. Microbiol.* **2000**, *66*, 4890–4896. [CrossRef] [PubMed]

97. Willford, J.; Mendonca, A.; Goodridge, L. Water Pressure Effectively Reduces Salmonella enterica Serovar Enteritidis on the Surface of Raw Almonds. *J. Food Prot.* **2008**, *4*, 825–829. [CrossRef]

98. Bello, E.F.T.; Martínez, G.G.; Klotz Ceberio, B.F.; Rodrigo, D.; López, A.M. High Pressure Treatment in Foods. *Foods* **2014**, *3*, 476–490. [CrossRef] [PubMed]

99. Black, E.P.; Setlow, P.; Hocking, A.D.; Stewart, C.M.; Kelly, A.L.; Hoover, D.G. Response of spores to high-pressure processing. *Compr. Rev. Food Sci. Food Saf.* **2007**, *6*, 103–119. [CrossRef]

100. Hao, H.; Zhou, T.; Koutchma, T.; Wu, F.; Warriner, K. High hydrostatic pressure assisted degradation of patulin in fruit and vegetable juice blends. *Food Control* **2016**, *62*, 237–242. [CrossRef]

101. Martínez-Rodríguez, Y.; Acosta-Muñiz, C.; Olivas, G.I.; Guerrero-Beltrán, J.; Rodrigo-Aliaga, D.; Mujica-Paz, H.; Welti-Chanes, J.; Sepulveda, D.R. Effect of high hydrostatic pressure on mycelial development, spore viability and enzyme activity of *Penicillium Roqueforti*. *Int. J. Food Microbiol.* **2014**, *168–169*, 42–46. [CrossRef] [PubMed]

102. Smith, K.; Mendonca, A.; Jung, S. Impact of high-pressure processing on microbial shelf-life and protein stability of refrigerated soymilk. *Food Microbiol.* **2009**, *26*, 794–800. [CrossRef] [PubMed]

103. Torres, J.A.; Saraiva, J.A.; Guerra-Rodríguez, E.; Aubourg, S.P.; Vázquez, M. Effect of combining high-pressure processing and frozen storage on the functional and sensory properties of horse mackerel (*Trachurus trachurus*). *Innov. Food Sci. Emerg. Technol.* **2014**, *21*, 2–11. [CrossRef]

foods

MDPI

Article

Can *Zymomonas mobilis* Substitute *Saccharomyces cerevisiae* in Cereal Dough Leavening?

Alida Musatti, Chiara Mapelli, Manuela Rollini, Roberto Foschino and Claudia Picozzi *

Dipartimento di Scienze per gli Alimenti, la Nutrizione, l'Ambiente, Università degli Studi di Milano, 20133 Milan, Italy; alida.musatti@unimi.it (A.M.); chiara.mapelli1@unimi.it (C.M.); manuela.rollini@unimi.it (M.R.); roberto.foschino@unimi.it (R.F.)
* Correspondence: claudia.picozzi@unimi.it; Tel.: +39-02-503-19174

Received: 28 March 2018; Accepted: 13 April 2018; Published: 16 April 2018

Abstract: Baker's yeast intolerance is rising among Western populations, where *Saccharomyces cerevisiae* is spread in fermented food and food components. *Zymomonas mobilis* is a bacterium commonly used in tropical areas to produce alcoholic beverages, and it has only rarely been considered for dough leavening probably because it only ferments glucose, fructose and sucrose, which are scarcely present in flour. However, through alcoholic fermentation, similarly to *S. cerevisiae*, it provides an equimolar mixture of ethanol and CO_2 that can rise a dough. Here, we propose *Z. mobilis* as a new leavening agent, as an alternative to *S. cerevisiae*, overcoming its technological limit with different strategies: (1) adding glucose to the dough formulation; and (2) exploiting the maltose hydrolytic activity of *Lactobacillus sanfranciscensis* associated with *Z. mobilis*. CO_2 production, dough volume increase, pH value, microbial counts, sugars consumption and ethanol production were monitored. Results suggest that glucose addition to the dough lets *Z. mobilis* efficiently leaven a dough, while glucose released by *L. sanfranciscensis* is not so well fermented by *Z. mobilis*, probably due to the strong acidification. Nevertheless, the use of *Z. mobilis* as a leavening agent could contribute to increasing the variety of baked goods alternative to those leavened by *S. cerevisiae*.

Keywords: *Zymomonas mobilis*; *Lactobacillus sanfranciscensis*; sourdough; dough leavening; bakery products; *Saccharomyces cerevisiae*; anti-*S. cerevisiae* antibodies

---

1. Introduction

In the last decades, the research in human nutrition has aimed both at improving food safety and demonstrating new healthy properties of foods or ingredients. Particularly in the grain cereals area, great attention has been paid to the study of sourdough microbial ecology [1,2] and to the positive effects of lactic acid bacteria (LAB) and yeast fermentation on the technological characteristics of dough. Therefore, several contributions have led to the enhancement of baked products by using a sourdough technology, which also matches consumer's choices in terms of their preference towards the valorisation of traditional products that can be certified [3]. Organic acid production impacts on sourdough texture and product shelf-life, with acetic acid also displaying anti-ropiness and antifungal activities [4,5]. Acidification also helps to activate endogenous cereal proteases that release peptides and amino acids related to flavour formation [6]. The production of bacteriocins allows microorganisms to control the sourdough ecosystem [7], while the synthesis of homo-polysaccharides delays firmness and staling [6]. Sourdough fermentation can also have positive nutritional implications by biodegrading phytates, thus increasing mineral bioavailability, and by lowering the glycaemic response to the consumption of baked goods [6].

Nevertheless, adverse food reactions, such as baker's yeast intolerance, have recently been increasing among Western population [8]. Apart from in well-known alcoholic beverages, such as beer, wine and cider, and baked goods, *S. cerevisiae* is also used in savoury spreads, as a food supplement in 'multi-vitamin' preparations and 'probiotics' in animal feed [9], and even in vaccine production [10]. It is therefore clear that we are often exposed to yeast parietal components [11].

Several studies report that an adverse response to baker's yeast occurs in a proportion of patients with Inflammatory Bowel Disease (IBD). In particular, in patients with Crohn's disease (CD), *S. cerevisiae* is recognized as an antigen, and anti-*S. cerevisiae* antibodies (ASCAs), directed against the cell wall mannan (phosphopeptidomannan) of yeast, have been identified as an important serological marker of this pathogenesis. However, the determination of ASCAs is also reliable in other autoimmune disorders besides CD [12]. Environmental factors such as food antigens may play an important role in the pathogenesis of autoimmune disorders [10,13] and obesity [9]. Although there is scarce literature on allergy-hypersensitivity to yeasts, some clinical conditions might benefit from reduced exposure to these microorganisms [14].

Based on these considerations, the study of new microbial resources to be applied in leavened goods may be considered of actual relevance; in this context, the possibility of replacing *S. cerevisiae* is noteworthy. The use of *Zymonomas mobilis* as leavening agent can contribute to an increase in the variety of bakery products alternative to those leavened by yeast in order to meet the specific demands of consumers. *Z. mobilis* can therefore be an interesting candidate to create a new food area of yeast-free baked goods. This bacterium is commonly used in tropical areas as a fermenting agent of plant saps to obtain alcoholic beverages such as pulque [15]. *Z. mobilis* ferments only glucose, fructose and sucrose, and through alcoholic fermentation it provides an equimolar mixture of ethanol and CO_2 that can theoretically leaven a dough [16], just like *S. cerevisiae*. The narrow range of fermentable substrates is a technological limit of *Z. mobilis* vs *S. cerevisiae* that may be overcome by: (1) adding a fermentable sugar to the dough formulation; or (2) exploiting maltose hydrolytic activity of *Lactobacillus sanfranciscensis* associated with *Z. mobilis*. This unconventional association has been investigated as a model system (higher cell concentration and leavening temperature, shorter leavening time) in a previous paper [17]. The present research aims to compare *Z. mobilis* leavening performance when glucose is added to the dough both with its fermentative ability when *Z. mobilis* is in association with *L. sanfranciscensis* and in doughs formulated and processed similarly to a type I sourdough.

2. Materials and Methods

2.1. Microorganisms and Maintenance

Z. mobilis subs. *mobilis* type strain DSM 424 (DSMZ: Deutsche Sammlung von Mikroorganismen und Zellkulturen GmbH) and *L. sanfranciscensis* DSM 20663 were used in this study.

Z. mobilis was maintained in liquid DSM medium, while biomass production was carried out in liquid IC G20 medium (as previously reported) [16]. Both media contain bacto-peptone (Costantino SpA, Turin, Italy) 10 g/L and glucose (Sigma Aldrich, St. Louis, MO, USA) 20 g/L, while they differ for yeast extract (Costantino SpA) 10 g/L present in DSM medium and of casein enzymatic hydrolysate (Costantino SpA) 10 g/L in IC G20. For both media, the pH was set at 6.8, and sterilization occurred at 112 °C for 30 min.

L. sanfranciscensis was maintained and cultivated in MRSm medium as reported elsewhere [17]. Cultures were incubated at 30 °C in stationary conditions for 16–24 h. Stock cultures of both microorganisms were stored at −80 °C in the same media (DSM for *Z. mobilis* and MRSm for *L. sanfranciscensis*) added with 20% (*v*/*v*) glycerol (VWR International, Leuven, Belgium).

2.2. Biomass Production

Z. mobilis was cultured in 1 L flasks containing 600 mL of liquid IC G20 medium, inoculated with 5% (*v*/*v*) of a 9 h pre-culture grown in DSM medium. *L. sanfranciscensis* was grown in 1 L flasks containing 600 mL of MRSm medium, inoculated with 2% (*v*/*v*) of a 24 h pre-grown culture in the same medium. Cultures were incubated at 30 °C in stationary conditions for 16 h for *Z. mobilis* and 24 h for *L. sanfranciscensis*.

The determination of the cell biomass was performed by spectrophotometric measurement (OD 600 nm, 6705 UV-Vis Spectrophotometer, Jenway, UK). For each strain, at 16 h in the case of *Z. mobilis* and 24 h for *L. sanfranciscensis*, a calibration curve was built (OD 600 vs. CFU (colony-forming unit)/mL) to determine the proper culture volume to add in the dough preparation (cell concentration expressed as Log CFU/g dough).

2.3. Dough Production and Analytical Determinations

Doughs were prepared with 333 g of a commercial type 0 Manitoba wheat flour (Simec SpA, Santa Giusta, Oristano, Italy) and 167 mL of distilled water, with or without addition of 1 or 5% (*w/w* flour) glucose. *Z. mobilis* was added alone (7 Log CFU/g dough) or with *L. sanfranciscensis* (5 Log CFU/g dough) yielding to 100:1 ratio *Zymomonas:Lactobacillus* cells. Ingredients were mixed in a food mixer (CNUM5ST, Bosch, Stuttgart, Germany) at speed 1 for 6 min. The dough was divided into 3 sections, treated as follows and then incubated at 26 °C:

- 400 g, inserted into a 1 L graduate cylinder to evaluate the dough volume increase up to 24 h of leavening;
- 25 g, inserted into a double chamber flask connected with a graduate burette filled with acidified water, to evaluate the total amount of CO_2 produced during leavening [16];
- The remaining sample was left to leaven into a Becker; samples were taken at appropriate intervals to determine dough pH, microbial counts and to carry out HPLC (high performance liquid chromatography) analysis.

Each analysis was performed at 0, 8, 16 and 24 h of leavening time.

2.4. Evaluation of Dough Volume Increase and Total CO_2 Production

The increase in the dough volume (mL) was evaluated at appropriate time intervals through the record of the level reached by the dough inside the graduate cylinder. CO_2 production (mL) was monitored by measuring the level reached by the liquid present inside the burette connected to the double chamber flask.

2.5. Determination of the Microbial Populations in Doughs

Approximately 10 g of dough sample were decimally diluted in sterile peptone water (10 g/L Bacto-peptone (Costantino SpA), pH 6.8) and homogenized in a Stomacher 400 Circulator (Seward, Worthing, UK) for 5 min at 260 rpm. The appropriate dilutions were plated onto MRSm agar (MRSm broth added with agar 15 g/L) for the determination of *L. sanfranciscensis* population, as well as onto DSM agar (DSM broth added with agar 15 g/L) for *Z. mobilis*. Plates were incubated at 30 °C for 3 d in anaerobic conditions. Aerobic bacterial count (ABC) was determined by pour plating in Tryptic Soy Agar (TSA, Scharlab, Barcelona, Spain) after incubation at 30 °C for 48–72 h. The enumeration of yeasts and moulds were carried out in Yeast Glucose Chloramphenicol Agar (YGC-Scharlab, Barcelona, Spain) plates after incubating at 25 °C for 3–5 day.

2.6. HPLC Analyses and pH Monitoring

Maltose and glucose consumption, as well as ethanol production during leavening, were measured through an HPLC system (L 7000, Merck Hitachi, Tokyo, Japan) as reported by Musatti et al. [17]. Briefly, 4 mL of homogenized dough samples were centrifuged (Eppendorf 5804 (Hamburg, Germany), 10,600× *g*, 10 min) and supernatants were filtered (0.45 µm syringe filter, VWR International, Radnor, PA, USA) before HPLC analysis. Data refer to 1 g dough (mg/g dough).

Dough pH was monitored at different intervals on the integral undiluted dough sample (pH-meter Eutech Instruments pH 510, Toronto, ON, Canada).

2.7. Statistical Analysis

All samples were prepared and analysed at least in triplicate. The effect of two factors, such as % glucose addition or *L. sanfranciscensis* co-inoculation, on some fermentation parameters were investigated by ANOVA according to the general linear model. Results of microbiological counts were transformed in the respective decimal logarithms to match a normal distribution of values. Data were processed with Statgraphic R Plus 5.1 for Windows (StatPoint, Inc., Herndon, VA, USA). When the effect was significant ($p < 0.05$), differences between means were separated by LSD test of multiple comparisons.

3. Results and Discussion

3.1. Trials with Glucose Addition into Dough

Dough samples were prepared with or without glucose addition (1% and 5% *w/w*). When glucose was not added, *Z. mobilis* fermented only the glucose amount naturally present in the flour (around 2.01 ± 0.59 mg/g dough). However, even if there are some hydrolytic enzymatic activities in the dough due to the presence of endogenous amylases, the low glucose concentration does not allow adequate CO_2 production to obtain a suitable dough volume increase by *Zymomonas*, especially in the first times of incubation.

The need to add a fermentable carbon source to the flour, in order to obtain a leavening of the dough, had already been highlighted in a previous work [16]. Actually, the results confirmed that the addition of glucose increases the CO_2 production ($p = 0.001$), and that in the three tested conditions mean values became statistically different at 16 h leavening time ($p = 0.007$) (Figure 1). Similarly, the addition of glucose allowed the doubling of the dough volume within the considered incubation time. As expected, the highest CO_2 production is related to the highest dough volume increase; in particular, with 5% glucose, the mean value of dough volume reached more than 850 mL ($p = 0.019$) with respect to an average of 815 or 735 mL with 1% or without glucose addition, respectively. CO_2 production was also related to bacterial growth; when no glucose is added, *Z. mobilis* grew approximately 1.3 Log CFU/g in 24 h, and around 1.6–2 Log CFU/g in the presence of 1% and 5% glucose, respectively. The performances obtained in dough samples with the two glucose concentrations were not significantly different between them, but both were statistically different from those obtained without glucose ($p < 0.001$).

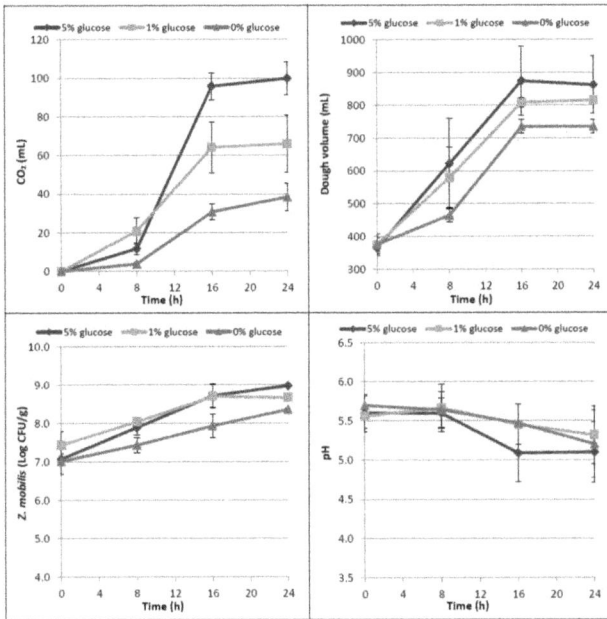

Figure 1. Time course of CO_2 production (mL), dough volume increase (mL), microbial growth of *Z. mobilis* (Log CFU/g) and dough pH in the three tested conditions (0%, 1%, 5% *w/w* glucose).

Results from HPLC analysis confirmed the increase of maltose during leavening time ($p < 0.001$) due to hydrolytic activity of the flour amylases and *Z. mobilis* inability to use this sugar (Table 1). At up to 8 h of incubation, the ethanol formation was not statistically different ($p = 0.414$), even if the three tested conditions had different levels of glucose. Then, glucose was mainly consumed between 8 and 16 h, producing CO_2 and ethanol in higher amounts in samples to which 5% glucose was added, as expected.

Table 1. Maltose, glucose and ethanol concentrations (expressed in terms of mg/g dough, mean and standard deviation (St. dev.)) present at 0, 8, 16 and 24 h in doughs leavened by *Z. mobilis* with 0%, 1%, 5% (*w/w*) of glucose added respect to the flour.

Glucose (% *w/w* flour)	Time (h)	Maltose (mg/g)		Glucose (mg/g)		Ethanol (mg/g)	
		Mean	St. dev.	Mean	St. dev.	Mean	St. dev.
0	0	10.50	0.96	1.96	0.06	0.00	0.00
	8	14.59	1.67	1.12	0.13	1.14	0.13
	16	17.71	3.16	0.23	0.33	2.79	0.49
	24	15.69	2.65	0.00	0.00	3.38	0.33
1	0	10.55	1.11	8.72	0.41	0.00	0.00
	8	15.48	5.22	3.42	0.62	1.84	0.41
	16	19.00	1.59	1.23	0.54	4.03	1.02
	24	21.43	4.49	0.92	0.44	3.96	0.77
5	0	8.33	0.95	35.72	2.95	0.00	0.00
	8	15.72	3.29	34.56	1.62	0.72	1.01
	16	18.57	2.84	2.66	1.48	9.73	0.13
	24	20.80	2.92	1.46	0.40	13.65	2.46

3.2. Bacterial Association Z. mobilis-L. sanfranciscensis

The association of *Zymomonas* with lactic acid bacteria has already been described in various food products, especially in some fermented drinks [18–21]. From this perspective, the possibility of obtaining a gradual glucose release in the dough exploiting the maltose hydrolytic activity of *Lactobacillus sanfranciscensis* was investigated [17].

When *L. sanfranciscensis* and *Z. mobilis* were inoculated together, the mean values of CO_2 production and dough volume increase did not significantly differ from those obtained with the use of *Z. mobilis* alone (Figure 2). Dealing with the single leavened samples, dough volumes at 16 and 24 h were found to be statistically different from those observed by using *L. sanfranciscensis* alone ($p = 0.023$ and 0.024, respectively). These results indicate that the contribution of the two microorganisms in association is not additive. Respect to the trials performed with glucose addition, in which the dough volumes nearly doubled in the first 8 h, the dough volume increased less than 20% in the case of the microbial association. As regards the trends of acidification and bacterial counts during the incubation time, the obtained data proved to be strongly affected by the presence of *L. sanfranciscensis*: the pH decreased to values of around 4 and the LAB growth (plus 4 Log CFU/g) was not influenced by the presence of *Zymomonas*. On the contrary, *Z. mobilis*, when grown in association with *L. sanfranciscensis*, statistically reduced ($p = 0.002$) its cell concentration from 16 h leavening onward. This behavior is probably due to the strong acidification of the medium produced by *L. sanfranciscensis*, able to affect both *Z. mobilis* vitality and fermentation ability.

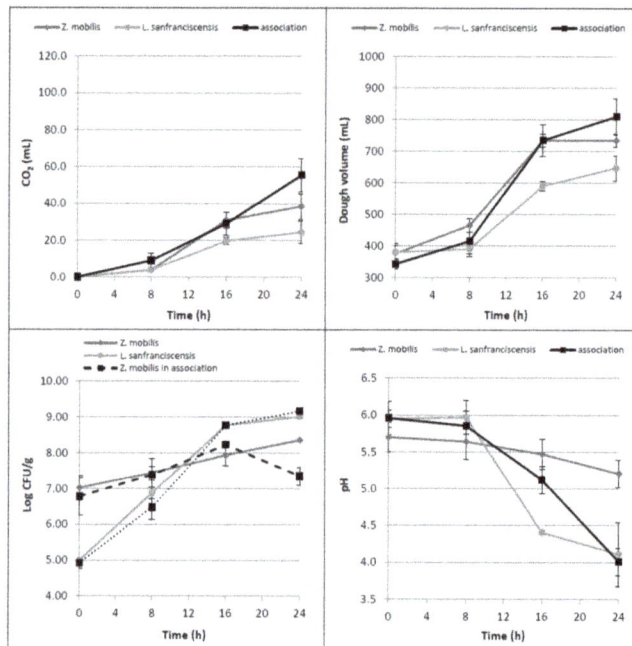

Figure 2. Time course of CO_2 production (mL), dough volume increase (mL), microbial growth (Log CFU/g) as well as dough pH, in doughs leavened by *Z. mobilis* and *L. sanfranciscensis* alone or by their association.

HPLC data confirmed that *L. sanfranciscensis* consumed maltose ($p = 0.012$) and released glucose ($p = 0.003$) in the dough [4], that it is not totally consumed by *Z. mobilis* (Table 2).

In summary, these results highlight that when inoculated alone, *Z. mobilis* is able to consume all the glucose present in a dough, while when coupled with *L. sanfranciscensis*, its fermentative performance decreases. Furthermore, the presence of *L. sanfranciscensis* did not lead to a significant ethanol yield increase, even if it can consume the available maltose in flour and release glucose.

Table 2. Maltose, glucose and ethanol concentrations (expressed in terms of mg/g dough, mean and standard deviation) present at 0, 8, 16 and 24 h in doughs leavened by *Z. mobilis*, *L. sanfranciscensis* and their association.

Microorganism	Time(h)	Maltose (mg/g)		Glucose (mg/g)		Ethanol (mg/g)	
		Mean	St. dev.	Mean	St. dev.	Mean	St. dev.
Lactobacillus sanfranciscensis (5 Log CFU/g)	0	11.33	1.46	2.08	1.01	0.00	0.00
	8	16.68	0.66	3.20	0.82	0.00	0.00
	16	9.43	0.60	4.01	0.28	0.00	0.00
	24	10.80	0.80	4.54	0.52	0.00	0.00
Zymomonas mobilis (7 Log CFU/g)	0	10.50	0.96	1.96	0.06	0.00	0.00
	8	14.59	1.67	1.12	0.13	1.14	0.13
	16	17.71	3.16	0.23	0.33	2.79	0.49
	24	15.69	2.65	0.00	0.00	3.38	0.33
L. sanfranciscensis coupled with *Z. mobilis* (5–7 Log CFU/g)	0	8.96	1.82	1.04	0.56	0.00	0.00
	8	14.21	2.70	0.92	0.31	0.75	0.28
	16	13.58	0.94	0.56	0.79	3.73	0.70
	24	12.33	2.76	0.82	0.35	4.91	1.23

4. Conclusions

The results obtained demonstrate that *Z. mobilis* is able to efficiently leaven a dough when glucose is present in the dough formulation. On the other hand, although the metabolic activities of LAB have positive effects on the structural and sensorial properties of the baked product [5,6], the traditional back-slopping sourdough technology [22] cannot be proposed due to the accelerated acidification of the dough impairing the growth of *Zymomonas*. In fact, preliminary trials have evidenced that, independently of the initial cell ratio between the two bacteria, *L. sanfranciscensis* always became the prevailing microbial population. This disproportion with *Z. mobilis* increased with refreshments, thus giving strongly acidified and poorly leavened doughs.

Future trials will be aimed at investigating microbial association with other LAB or the use of dough formulations naturally enriched of sugars fermentable by *Z. mobilis*. In this context, sucrose can also be considered an interesting alternative to glucose; nevertheless, the strain leavening performance with this carbon source has to be evaluated.

Acknowledgments: This work was founded by Bando Linea R&S per Aggregazioni, Regione Lombardia, Programma Operativo Regionale 2014–2020, Strategia "InnovaLombardia" (D.G.R. No. 2448/2014) Project number 145007.

Author Contributions: A.M. and C.P. conceived and designed the experiments; A.M. and C.M. performed the experiments; R.F. analysed the data; R.F. contributed reagents/materials/analysis tools; M.R. and C.P. wrote the paper.

Conflicts of Interest: The authors declare no conflict of interest. The founding sponsors had no role in the design of the study; in the collection, analyses, or interpretation of data; in the writing of the manuscript, and in the decision to publish the results.

References

1. Gobbetti, M. The sourdough microflora: Interactions of lactic acid bacteria and yeasts. *Food Sci. Technol.* **1998**, *9*, 267–274. [CrossRef]
2. De Vuyst, L.; Vrancken, G.; Ravyts, F.; Rimaux, T.; Weckx, S. Biodiversity, ecological determinants, and metabolic exploitation of sourdough microbiota. *Food Microbiol.* **2009**, *26*, 666–675. [CrossRef] [PubMed]

3. Picozzi, C.; D'Anchise, F.; Foschino, R. PCR detection of *Lactobacillus sanfranciscensis* in sourdough and Panettone baked product. *Eur. Food Res. Technol.* **2006**, *222*, 330–335. [CrossRef]

4. Gobbetti, M.; De Angelis, M.; Corsetti, A.; Di Cagno, R. Biochemistry and physiology of sourdough lactic acid bacteria. *Trends Food Sci. Technol.* **2005**, *16*, 57–69. [CrossRef]

5. Arendt, E.K.; Ryan, L.A.M.; Bello, F.D. Impact of sourdough on the texture of bread. *Food Microbiol.* **2007**, *24*, 165–174. [CrossRef] [PubMed]

6. Angioloni, A.; Romani, S.; Gaetano Pinnavaia, G.; Dalla Rosa, M. Characteristics of bread making doughs: Influence of sourdough fermentation on the fundamental rheological properties. *Eur. Food Res. Technol.* **2006**, *222*, 54–57. [CrossRef]

7. Messens, W.; De Vuyst, L. Inhibitory substances produced by Lactobacilli isolated from sourdoughs—A review. *Int. J. Food Microbiol.* **2002**, *72*, 31–43. [CrossRef]

8. Mansueto, P.; Montalto, G.; Pacor, M.L.; Esposito-Pellitteri, M.; Ditta, V.; Lo Bianco, C.; Leto-Barone, S.M.; Di Lorenzo, G. Food allergy in gastroenterologic diseases: Review of literature. *World J. Gastroenterol.* **2006**, *12*, 7744–7752. [CrossRef] [PubMed]

9. Salamati, S.; Martins, C.; Kulseng, B. Baker's yeast (*Saccharomyces cerevisiae*) antigen in obese and normal weight subjects. *Clin. Obes.* **2015**, *5*, 42–47. [CrossRef] [PubMed]

10. Rinaldi, M.; Perricone, R.; Blank, M.; Perricone, C.; Shoenfeld, Y. Anti-*Saccharomyces cerevisiae* autoantibodies in autoimmune diseases: From bread baking to autoimmunity. *Clin. Rev. Allergy Immunol.* **2013**, *45*, 152–161. [CrossRef] [PubMed]

11. Sicard, D.; Legras, J.-L. Bread, beer and wine: Yeast domestication in the *Saccharomyces sensu stricto* complex. *Comptes Rendus Biol.* **2011**, *334*, 229–236. [CrossRef] [PubMed]

12. Israeli, E.; Grotto, I.; Gilburd, B.; Balicer, R.D.; Goldin, E.; Wiik, A.; Shoenfeld, Y. Anti-*Saccharomyces cerevisiae* and antineutrophil cytoplasmic antibodies as predictors of inflammatory bowel disease. *Gut* **2005**, *54*, 1232–1236. [CrossRef] [PubMed]

13. Muratori, P.; Muratori, P.; Muratori, L.; Guidi, M.; Maccariello, S.; Pappas, G.; Ferrari, R.; Gionchetti, P.; Campieri, M.; Bianchi, F.B. Anti-Saccharomyces cerevisiae antibodies (ASCA) and autoimmune liver diseases. *Clin. Exp. Immunol.* **2003**, *132*, 473–476. [CrossRef] [PubMed]

14. Bansal, R.A.; Tadros, S.; Bansal, A.S. Beer, Cider, and Wine Allergy. *Case Rep. Immunol.* **2017**, *2017*, 7958924. [CrossRef] [PubMed]

15. Sahm, H.; Bringer-Meyer, S.; Sprenger, G.A. Proteobacteria: Alpha and Beta Subclasses. In *The Prokaryotes*, 3rd ed.; Dworkin, M., Falkow, S., Rosenberg, E., Schleifer, K.-H., Stackebrandt, E., Eds.; Springer: Berlin, Germany, 2006; Volume 5, ISBN 978-0-387-25476-0.

16. Musatti, A.; Rollini, M.; Sambusiti, C.; Manzoni, M. *Zymomonas mobilis*: Biomass production and use as a dough leavening agent. *Ann. Microbiol.* **2015**, *65*, 1583–1589. [CrossRef]

17. Musatti, A.; Mapelli, C.; Foschino, R.; Picozzi, C.; Rollini, M. Unconventional bacterial association for dough leavening. *Int. J. Food Microbiol.* **2016**, *237*, 28–34. [CrossRef] [PubMed]

18. Alcántara-Hernández, R.J.; Rodríguez-Álvarez, J.A.; Valenzuela-Encinas, C.; Gutiérrez-Miceli, F.A.; Castañón-González, H.; Marsch, R.; Ayora-Talavera, T.; Dendooven, L. The bacterial community in "taberna" a traditional beverage of Southern Mexico. *Lett. Appl. Microbiol.* **2010**, *51*, 558–563. [CrossRef] [PubMed]

19. Escalante, A.; Giles-Gómez, M.; Hernández, G.; Córdova-Aguilar, M.S.; López-Munguía, A.; Gosset, G.; Bolívar, F. Analysis of bacterial community during the fermentation of pulque, a traditional Mexican alcoholic beverage, using a polyphasic approach. *Int. J. Food Microbiol.* **2008**, *124*, 126–134. [CrossRef] [PubMed]

20. Nwachukwu, I.N.; Ibekwe, V.I.; Anyanwu, B.N. Investigation of some physicochemical and microbial succession parameters of palm wine. *J. Food Technol.* **2006**, *4*, 308–312.

21. Valadez-Blanco, R.; Bravo-Villa, G.; Santos-Sánchez, N.F.; Velasco-Almendarez, S.I.; Montville, T.J. The Artisanal Production of Pulque, a Traditional Beverage of the Mexican Highlands. *Probiotics Antimicrob. Proteins* **2012**, *4*, 140–144. [CrossRef] [PubMed]

22. Vogelmann, S.A.; Hertel, C. Impact of ecological factors on the stability of microbial associations in sourdough fermentation. *Food Microbiol.* **2011**, *28*, 583–589. [CrossRef] [PubMed]

foods

MDPI

Article

Composition, Protein Profile and Rheological Properties of Pseudocereal-Based Protein-Rich Ingredients

Loreto Alonso-Miravalles and James A. O'Mahony *

School of Food and Nutritional Sciences, University College Cork, Cork T12 Y337, Ireland;
116221127@umail.ucc.ie
* Correspondence: sa.omahony@ucc.ie; Tel.: +353-21-490-3625

Received: 31 March 2018; Accepted: 26 April 2018; Published: 7 May 2018

Abstract: The objectives of this study were to investigate the nutrient composition, protein profile, morphology, and pasting properties of protein-rich pseudocereal ingredients (quinoa, amaranth, and buckwheat) and compare them to the more common rice and maize flours. Literature concerning protein-rich pseudocereal ingredients is very limited, mainly to protein profiling. The concentrations of macronutrients (i.e., ash, fat, and protein, as well as soluble, insoluble and total dietary fibre) were significantly higher for the protein-rich variants of pseudocereal-based flours than their regular protein content variants and the rice and maize flours. On profiling the protein component using sodium dodecyl sulfate–polyacrylamide gel electrophoresis (SDS-PAGE), all samples showed common bands at ~50 kDa and low molecular weight bands corresponding to the globulin fraction (~50 kDa) and albumin fraction (~10 kDa), respectively; except rice, in which the main protein was glutelin. The morphology of the starch granules was studied using scanning electron microscopy with quinoa and amaranth showing the smallest sized granules, while buckwheat, rice, and maize had the largest starch granules. The pasting properties of the ingredients were generally similar, except for buckwheat and amaranth, which showed the highest and lowest final viscosity, respectively. The results obtained in this study can be used to better understand the functionality and food applications of protein-rich pseudocereal ingredients.

Keywords: pseudocereal; cereal; protein-rich ingredients; macronutrient; protein profile; morphology; rheological properties

1. Introduction

The global protein demand for the 7.3 billion inhabitants of the world is approximately 202 million tonnes annually [1]. The expected continuous growth of the global population to 9.6 billion people by 2050 is creating an ever-greater need to identify and develop sustainable solutions for provision of high-quality food protein [2,3]. Plant-based protein ingredients are becoming more popular due to their contribution to environmental sustainability and to food security challenges, in addition to their cost-effectiveness, compared with animal-based proteins [4]. However, replacing animal-based protein ingredients with plant-origin material is not easy due mainly to important differences in composition and taste/flavour [5]. Moreover, applications of plant proteins are poorly studied and commercially limited due mainly to their techno-functional properties (e.g., poor solubility), anti-nutritional components, off-flavour, and colour [6,7].

Quinoa, amaranth, and buckwheat are non-conventional sources of protein that have been the subject of limited studies in recent years, although their cultivation goes back thousands of years [8,9]. They are gluten-free dicotyledonous grains, referred to as pseudocereals, with somewhat similar composition and nutritional value to cereals, such as rice and maize [10,11]. Quinoa and amaranth are cultivated in South America, and buckwheat, originally from Central Asia, is now also cultivated in Central and Eastern Europe [12]. Their main compositional component is starch [13] which forms semi-crystalline structures referred to as "starch granules", and depending on the botanical source, these granules vary in size, shape, and amylose:amylopectin ratio [14], which consequently influences the techno-functional properties of the flour ingredients [15,16]. Protein, fibre, fat, minerals, and vitamins are the remaining macro- and micro-nutrients that constitute pseudocereals [9,17]. The protein content of amaranth, buckwheat, and quinoa, has been reported to be 12.0%–18.9% and the concentrations of essential amino acids, particularly, cysteine and methionine, are known to be higher than in some common cereals such as rice and maize [12].

Regarding classification of pseudocereal proteins, the literature in this area is often inconsistent and contradictory [17]. Several authors [18,19] have reported globulins and albumins to be the main proteins in quinoa, amaranth, and buckwheat, in contrast to other cereals, such as rice, where the main proteins are glutelin and prolamins [20]. Amaranth, quinoa, and buckwheat are also good sources of dietary fibre, which has proven effects in promoting desirable physiological outcomes, such as lowering blood cholesterol and increased satiety, due to its resistance to digestion and absorption in the small intestine, followed by complete or partial fermentation in the large intestine [21–23]. In addition, pseudocereals are rich in micro-nutrients such as calcium, magnesium, and iron and good sources of vitamin E and riboflavin [24].

These macro- and micro-nutrients are located in different parts of the grain (Figure 1). In amaranth and quinoa seeds, the embryo or germ, which is circular in shape, surrounds the starch-rich perisperm, and together with the seed coat, represent the bran fraction, which is relatively rich in fat and protein [25]. In contrast, in buckwheat seeds, starch reserves are stored in the endosperm, as in common cereals, and the embryo, rich in fat and protein, extends through the starchy endosperm [26]. Protein-enriched fractions can be prepared from such pseudocereal grains using two principal approaches—dry or wet fractionation techniques [27]. Dry fractionation employs mechanical forces (milling and air/size classification) and is a more sustainable means of obtaining protein-rich fractions, while wet fractionation techniques use large quantities of water, chemicals (e.g., for pH adjustment), and a final drying step that consumes energy [4,28]. Therefore, protein-rich fractions from pseudocereals can offer unique nutritional and technological properties that have not yet been fully investigated or tested in food applications [29–31].

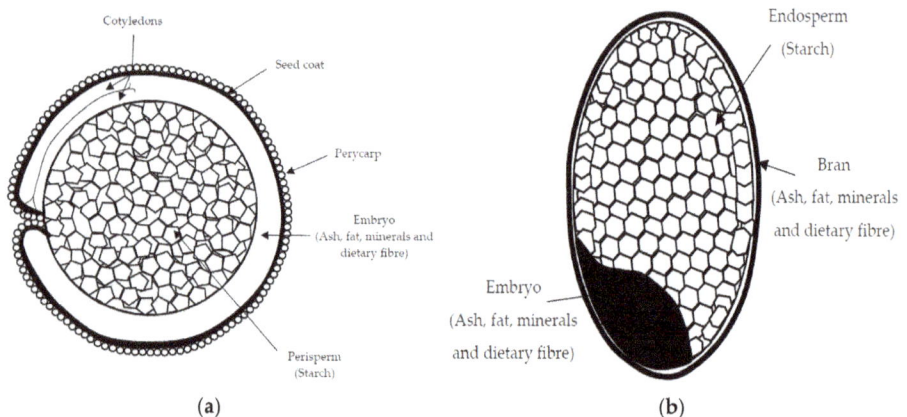

Figure 1. *Cont.*

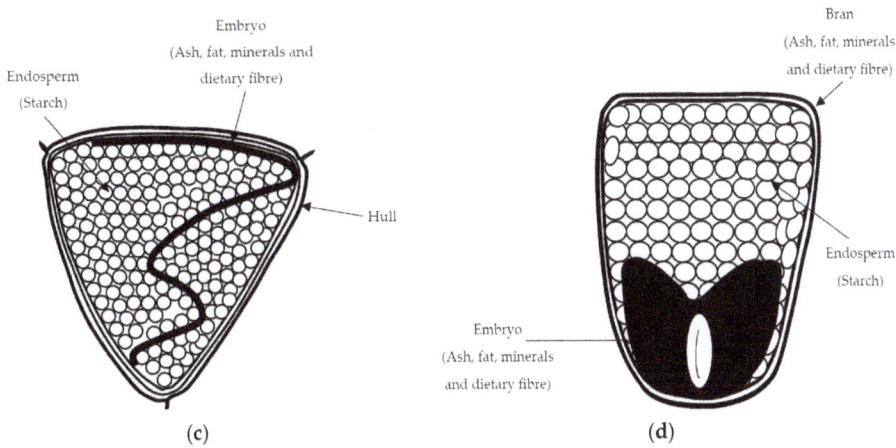

Figure 1. Schematic representation of grain structure of quinoa and amaranth (a); rice (b); buckwheat (c); and maize (d).

The aim of this work was to determine systematically the nutritional composition, protein profile, and physical properties of several novel protein-enriched ingredients from quinoa, amaranth, and buckwheat and compare them to regular protein content pseudocereal and cereal flours. These protein-rich fractions have great potential as ingredients, not only for their nutritional value (e.g., rich in protein and fibre) but also for their technological functionality (e.g., starch pasting properties). Scientific information on pseudocereal protein-rich fractions is scarce in the literature, thus, the results of this original and novel study can help with our understanding of the potential applications of these plant-based protein-rich ingredients in food formulations.

2. Materials and Methods

2.1. Cereal and Pseudocereal Flour Ingredients

Ten different regular and protein-rich cereal and pseudocereal flours/ingredients were analysed in this study. Seven of the flours were of pseudocereal origin: quinoa wholegrain flour (QWGF), quinoa dehulled flour (QDF), quinoa protein-rich flour (QPRF), amaranth wholegrain flour (AWGF), amaranth protein-rich flour (APRF), buckwheat dehulled flour (BDF), and buckwheat protein-rich flour (BPRF). Protein enrichment in the protein-rich flours was achieved using a dry milling approach. In brief, the grains were milled using either an impact or a jet mill, with different screen inserts used to produce flour and seed fragments; only buckwheat was milled using a jet mill. All grains, except amaranth, were sourced from commercial suppliers and had been de-hulled prior to milling. After milling, the protein-rich fractions were separated from the milled flours using size-based dry sieve classification. Rice flour (RF), rice protein concentrate (RPC), and maize flour (MF) were included in the study as comparator flour ingredients and were of cereal origin. All of the pseudocereal flours were provided by the Fraunhofer Institut (Munich, Germany) except the QWGF, which was purchased from Ziegler & Co. (Wunsiedel, Germany). The RF and RPC ingredients were purchased from Beneo (Tienen, Belgium) and the MF was purchased from the Quay Co-op (Cork, Ireland).

2.2. Chemical Composition

Moisture, ash, fat, and protein contents of samples were determined according to the standard methods of the Association of Analytical Chemists [32]. Moisture was determined by oven drying at 103 °C for 5 h (AOAC 925.10). The ash content was analysed by dry ashing in a muffle furnace at

500 °C for 5 h (AOAC 923.03). Fat determination was carried out following AOAC 922.06, using a Soxtec 2055 (Foss, Ballymount, Co., Dublin, Ireland). Total nitrogen content was determined by the Kjeldahl method (AOAC 930.29) using the following nitrogen-to-protein conversion factors: 6.25 for quinoa, buckwheat, and maize [12,33], 5.85 for amaranth [24], and 5.95 for rice ingredients [12]. Total carbohydrate was calculated by difference (i.e., 100—sum of protein, fat, ash, and moisture). Total starch (AOAC Methods 996.11 and AACC Method 76-13.01), damaged starch as a % of total starch (AACC method 76-31.01 and ICC method No. 164), and soluble (SDF), insoluble (IDF), and total dietary fibre (TDF) (AOAC Method 991.43 and AACC Method 32-07.01) contents were determined using enzyme kits (Megazyme, Bray, Co., Wicklow, Ireland). β-glucan, casein, and high-amylose maize starch were used as controls in dietary fibre analysis (K-TDFC; Megazyme, Wicklow, Ireland).

2.3. Electrophoretic Protein Profile Analysis

The protein profile was assessed by sodium dodecyl sulphate-polyacrylamide gel electrophoresis (SDS-PAGE) using precast gels (Mini-PROTEAN TGX, Bio-Rad Laboratories, Hercules, CA, USA) under non-reducing (method **I** and **II**) and reducing conditions (method **III**). The sample loading buffer contained 65.8 mM Tris-HCl (pH 6.8), 26.3% glycerol, 2.1% sodium dodecyl sulfate (SDS) and 0.01% bromophenol blue. The running buffer (10× Tris/Glycine/SDS, Bio-Rad Laboratories, Hercules, CA, USA) had a composition of 25 mM Tris, 192 mM glycine, and 0.1% SDS (*w/v*), pH 8.3. The staining solution used was Coomassie Brilliant Blue R-250 (Bio-Rad Laboratories, Hercules, CA, USA). The target final protein concentration was, in all cases, 1 mg/mL, and 10 μL of sample solution loaded into each well of the gel. For the preparation of the samples, three different methods were used. For method **I**, the approach of Abugoch et al. [34] was followed, with slight modifications. Briefly, the powder samples were mixed directly with the sample loading buffer at a concentration of 1 mg/mL, vortexed for 1 min until the powder was fully suspended and mixed over 2 h at 20 °C and at 250 rpm. For methods **II** and **III**, the approach of Amagliani et al. [20] was followed, with the modification that the powders were mixed with the protein extracting buffer overnight, and 1,4-dithiothreitol (DTT; 1%) was used in method **III** as a reducing agent.

2.4. Microstructural Analysis

The powders were mounted on aluminium stubs using double-sided adhesive carbon tape, and sputter coated with a 5 nm layer of gold/palladium (Au:Pd = 80:20) using a Quorum Q150R ES Sputter Coating Unit (Quorum Technologies Ltd., Sussex, UK). Subsequently, the samples were loaded into a sample tube and examined using a JSM-5510 scanning electron microscope (JEOL Ltd., Tokyo, Japan), operated at an accelerating voltage of 5 kV.

2.5. Pasting Behaviour

Pasting properties were studied using an AR-G2 controlled-stress rheometer equipped with a starch pasting cell (AR-G2; TA Instruments Ltd., Waters LLC, Leatherhead, UK). The internal diameter of the cell was 36.0 mm, the diameter of the rotor was 32.4 mm, and the gap between the two elements at the geometry base was 0.55 mm. A heating and cooling cycle described by Li et al. [35] was applied to 16% (*w/w*) suspensions of flours ingredients at a fixed shear rate of 17 rad/s.

2.6. Statistical Analysis

All the analyses were conducted in triplicate. The data generated was subjected to one-way analysis of variance (ANOVA) using R i386 version 3.3.1 (R foundation for statistical computing, Vienna, Austria). A Tukey's paired comparison test was used to determine statistically significant differences ($p < 0.05$) between mean values for different samples, at a 95% confidence level.

3. Results and Discussion

3.1. Chemical Composition

The dry matter that remains after moisture removal is commonly referred to as total solids [36]. Protein-rich samples had higher total solids ($p < 0.05$) content than their regular flour counterparts. The higher total solids content of these protein-rich ingredients can be an advantage from a microbiological and chemical stability perspective [37]. Ash refers to substances resulting from the incineration of dry matter in a powder sample and is directly related to the mineral content of the sample [38]. The protein-rich ingredients, QPRF, APRF, BPRF, and RPC showed higher ash contents (3.6%, 6.9%, 3.0%, and 3.4%, respectively) than the regular flours QWGF, QDF, AWGF, BDF, RF, and MF (2.3%, 1.8%, 2.4%, 1.5%, 0.8%, and 0.7%, respectively). Protein-rich flours are usually produced using dry fractionation approaches [28], classifying the parts of the grain that are rich in protein (e.g., embryo fraction) which results in a concomitant increase in other components such as minerals [5,25,39]. These pseudocereal protein-rich fractions with higher ash content would be expected to be enriched in selected minerals such as phosphorus, magnesium, and potassium that are located in embryonic tissues [33,40].

The fat content of the protein-rich ingredients QPRF, APRF, and BPRF (12.8%, 16.6%, and 4.8%, respectively) was significantly higher ($p < 0.05$) than the regular flours (Table 1). The higher fat content of the protein-rich ingredients was expected taking into consideration that the dry fractionation process classifies fractions rich in fat along with protein. Arendt and Zannini [40], reported that in quinoa, 49% of the total fat content is located in the embryo. Gamel et al. [41], reported 45% higher fat content in amaranth protein-rich flours, in comparison with a regular flour, and related it with the association of fat with cell wall materials and protein bodies during the protein enrichment process. BPRF showed the lowest value for fat (4.7%) among the protein-rich ingredients. In this study, the low fat content of BDF is most likely due to its relatively low level of protein enrichment (20%) which suggests lower enrichment in the embryo fraction where most of the fat is located. Also, Alvarez-Jubete et al. [26] stated that the fat content in quinoa and amaranth is two to three times higher than in buckwheat and common cereals. The fact that these pseudocereals have high levels of fat reduce the need for adding fat when these protein-rich flours are used as ingredients (e.g., baked products) where fat plays an important role in texture and flavour [41].

The protein-rich flour ingredients, QPRF, APRF, and BPRF, had values for protein of 33.3%, 38.6%, and 20.5%, respectively. The protein contents for pseudocereal flours ranged from 13.1% to 15.7% which are higher than the protein values for RF (8.2%) and MF (6.4%). These values are in accordance with the study of Mota et al. [12], who reported a protein content for pseudocereals significantly higher than in common cereals such as rice and maize. Moreover, a recent review by Navruz et al. [42] reported the nutritional and health benefits of quinoa, such as protein digestibility values similar to casein and higher lysine levels than other grains.

The values for starch in protein-rich samples were lower (21.4–47.3%) than those for the regular flours (50.5–61.6%). The lower values for starch in the protein-rich ingredients were expected as protein-rich ingredients are more enriched in the embryo fraction (rich in proteins), while the perisperm (quinoa and amaranth) or endosperm (buckwheat) where the starch granules are located, are less abundant. The level of starch damage is related to the process and the conditions (e.g., pressure or shear) used to obtain the protein-rich flour ingredients [43]. Such damage changes the granular structure of starch and influences the rheological and functional properties of the starch granules by modulating their water sorption and swelling capacity [43]. QWGF, QDF, QPRF, AWGF, RF, and MF showed similar levels of damaged starch (~7–12% of total starch); while APRF, BDF, and BPRF had the lowest levels of starch damage (~2%) (Table 1). The differences in damaged starch between the samples are usually related to the severity of the extraction process employed [20]. RPC showed the highest damaged starch content (88.3%), which might have arisen from the use of chemicals and aggressive environmental conditions (temperature and pH) in obtaining high protein levels in the final product [4].

Table 1. Macronutrient composition of quinoa wholegrain flour (QWGF), quinoa dehulled flour (QDF), quinoa protein-rich flour (QPRF), amaranth wholegrain flour (AWGF), amaranth protein-rich flour (APRF), buckwheat dehulled flour (BDF), buckwheat protein-rich flour (BPRF), rice flour (RF), rice protein concentrate (RPC), and maize flour (MF). Total dietary fibre (TDF). Values are means ± standard deviations of data from triplicate analysis.

	Moisture	Ash	Protein (% w/w)	Fat	Carbohydrate	Starch	Damaged Starch (% Total Starch)	TDF (% w/w)
Quinoa								
QWGF	9.01 ± 0.10 [d]	2.30 ± 0.00 [c]	13.1 ± 0.10 [b]	6.54 ± 0.07 [d]	69.0 ± 0.27 [d]	60.0 ± 2.58 [d]	10.6 ± 0.47 [d]	11.4 ± 1.10 [b]
QDF	8.86 ± 0.25 [d]	1.80 ± 0.10 [b]	15.7 ± 0.30 [b]	5.36 ± 0.61 [d]	68.3 ± 1.26 [d]	50.5 ± 1.40 [bc]	11.7 ± 0.32 [e]	9.75 ± 1.17 [b]
QPRF	5.25 ± 0.25 [a]	3.60 ± 0.19 [e]	33.3 ± 1.10 [d]	12.8 ± 0.73 [e]	45.0 ± 2.27 [b]	21.4 ± 0.81 [a]	10.4 ± 0.40 [d]	18.8 ± 0.23 [c]
Amaranth								
AWGF	8.94 ± 0.05 [d]	2.40 ± 0.02 [c]	14.6 ± 0.30 [b]	6.04 ± 0.10 [d]	68.1 ± 0.47 [d]	52.8 ± 1.45 [c]	12.2 ± 0.35 [e]	11.3 ± 0.86 [b]
APRF	7.76 ± 0.12 [b]	6.86 ± 0.18 [f]	38.6 ± 1.74 [e]	16.6 ± 0.08 [f]	30.2 ± 2.12 [a]	20.3 ± 0.31 [a]	2.61 ± 0.01 [b]	24.0 ± 2.56 [d]
Buckwheat								
BDF	8.75 ± 0.11 [d]	1.51 ± 0.31 [b]	14.2 ± 0.06 [b]	2.77 ± 0.05 [bc]	72.8 ± 0.53 [e]	61.6 ± 0.12 [d]	1.52 ± 0.06 [a]	10.3 ± 1.72 [b]
BPRF	6.86 ± 0.17 [c]	3.05 ± 0.10 [d]	20.5 ± 0.90 [c]	4.76 ± 0.15 [cd]	64.8 ± 1.32 [c]	47.3 ± 1.20 [b]	2.22 ± 0.07 [ab]	19.0 ± 0.48 [c]
Rice								
RF	8.89 ± 0.19 [d]	0.85 ± 0.05 [a]	8.22 ± 0.14 [a]	0.71 ± 0.08 [a]	81.3 ± 0.46 [f]	78.5 ± 0.82 [e]	10.7 ± 0.14 [f]	1.12 ± 0.20 [a]
RPC	6.24 ± 0.08 [a]	3.42 ± 0.24 [d]	75.0 ± 0.38 [f]	0.79 ± 0.00 [a]	14.6 ± 0.7 [g]	6.50 ± 0.71 [f]	88.3 ± 0.11 [g]	5.83 ± 0.41 [e]
Maize								
MF	12.2 ± 0.31 [e]	0.74 ± 0.04 [a]	6.42 ± 0.21 [a]	1.66 ± 0.02 [ab]	79.0 ± 0.58 [f]	76.0 ± 2.26 [e]	7.21 ± 0.25 [c]	2.00 ± 0.40 [a]

Values followed by different superscript letters (a–f) in the same column are significantly different (*p* < 0.05).

Dietary fibre denotes carbohydrate polymers which are not hydrolysed by the endogenous enzymes in the small intestine of humans [21,22]. Total dietary fibre (TDF) is divided into two categories, based on differences in solubility in water: soluble (SDF) and insoluble (IDF) dietary fibre. Protein-rich cereal ingredients showed significantly higher levels (19–24%) ($p < 0.05$) of TDF than the regular protein containing ingredients (1.1–11.5%) (Table 1 and Figure 2). Among the pseudocereal flours there were no significant differences ($p < 0.05$) in TDF, but they showed higher contents of TDF ($p < 0.05$) in comparison with RF and MF. These results were expected on comparison with literature data: Nascimento et al. [33], reported that pseudocereals can have seven times more fibre than common grains such as rice. The TDF values were similar to those found in other studies for quinoa [33,44,45] where values for TDF of 10.4%, 11.7%, and 12.7%, respectively, were reported, whereas Alvarez-Jubete et al. [26] reported slightly higher values for TDF (14.2%). The value for AWGF is in line with Nascimento et al. [33] who reported a TDF content of 11.3% for amaranth. Other authors, such as Repo-Carrasco et al. [46], reported slightly higher values (ranging from 14% to 16%) for amaranth (*Amaranthus caudatus*) flours. Regarding the soluble and insoluble dietary fibre fractions, the IDF fraction was higher than the SDF fraction in all the ingredients except for RF. This is in accordance with values reported in the literature for quinoa [47,48] and amaranth [19]. However, the IDF content of AWGF was slightly lower than that reported previously by Repo-Carrasco et al. [46] for the varieties Oscar Blanco (12.15%) and Centenario (13.92%). RPC had the lowest values for TDF, SDF, and IDF, which might be explained by the higher protein enrichment levels for this sample, which was in turn, associated with lower levels of other components such as starch, fat, and dietary fibre.

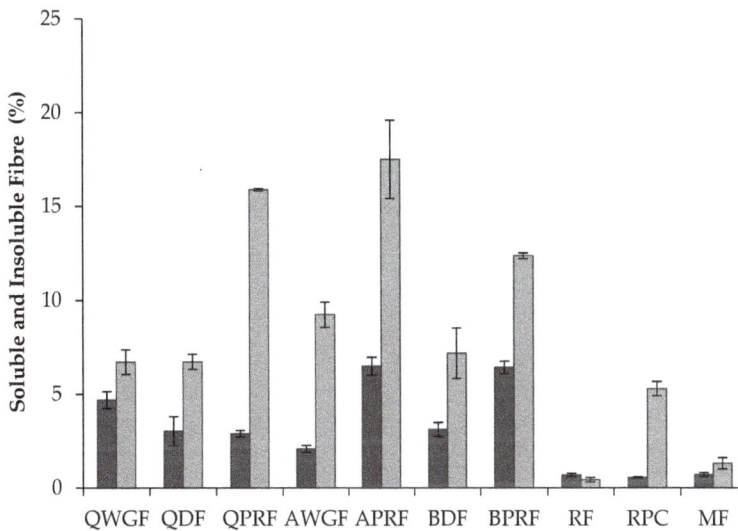

Figure 2. Soluble (■) and insoluble (□) dietary fibre content (% *w/w*) of quinoa wholegrain flour (QWGF), quinoa dehulled flour (QDF), quinoa protein-rich flour (QPRF), amaranth wholegrain flour (AWGF), amaranth protein-rich flour (APRF), buckwheat dehulled flour (BDF), buckwheat protein-rich flour (BPRF), rice flour (RF), rice protein concentrate (RPC), and maize flour (MF).

3.2. Protein Profile by SDS-PAGE Electrophoresis

SDS-PAGE analyses under non-reducing conditions (Figure 3a,b) and reducing conditions (Figure 3c) were performed using methods **I**, **II**, and **III**, respectively, as outlined in Section 2.3. All samples, except maize, showed common protein bands at ~50 kDa under non-reducing conditions (Figure 3a,b). This band corresponds to the globulin and glutelin fraction in pseudocereals and rice, respectively. For quinoa samples (QWGF, QDF, and QPRF), bands at ~50 kDa (Figure 3a,b) correspond to the 11S globulin fraction, also commonly referred to as chenopodin. Chenopodin consists of ~49 and 57 kDa subunits that are associated into a hexamer by non-covalent interactions [18,49]. When quinoa proteins are treated directly with the sample loading buffer (Figure 3a), two bands with molecular weight (MW) lower and higher than ~50 kDa can be observed. The higher intensity of the lower MW band (Figure 3a,b), suggests that this subunit is predominant in chenopodin protein. When the sample was treated with the protein extracting buffer containing SDS, urea, and thiourea (i.e., under non-reducing conditions; method **II** and Figure 3b), the chenopodin (~50 kDa) did not dissociate into bands of lower MW suggesting that disulphide bonds are the principal linkage between the subunits. In a similar manner to quinoa, the amaranth samples (AWGF and APRF), showed a band at ~50 kDa (Figure 3a,b), which corresponds to the hexameric 11S globulin or amarantin [17]. This major band might also be attributed to another glutelin-type protein which has similar molecular characteristics to those of amaranth 11S globulin [50]. Buckwheat samples, showed a main band at ~50 kDa, which may correspond to the major storage protein of buckwheat, the 13S legume-like globulin, and the minor storage protein, the trimer 8S vicilin-like globulin [51]. Rice samples also showed a major band at ~50 kDa (Figure 3a,b) which corresponds to the glutelin precursor [20].

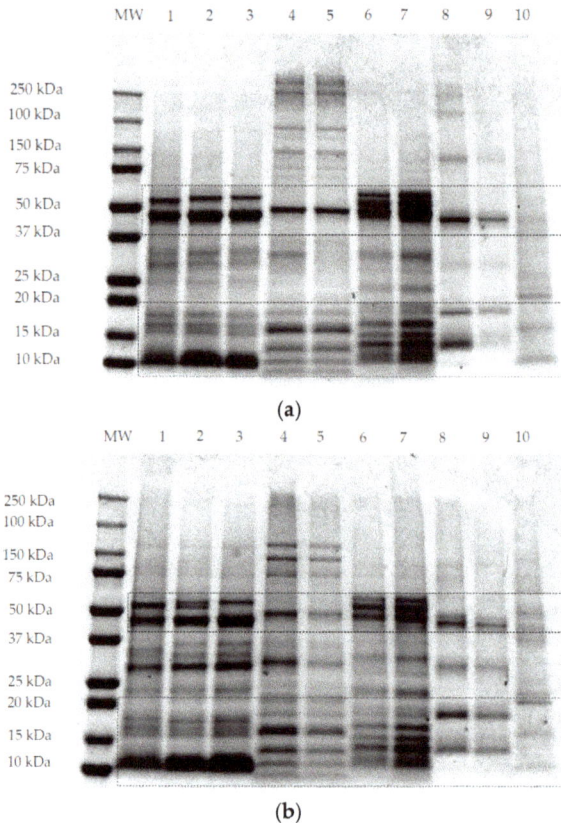

(a)

(b)

Figure 3. *Cont.*

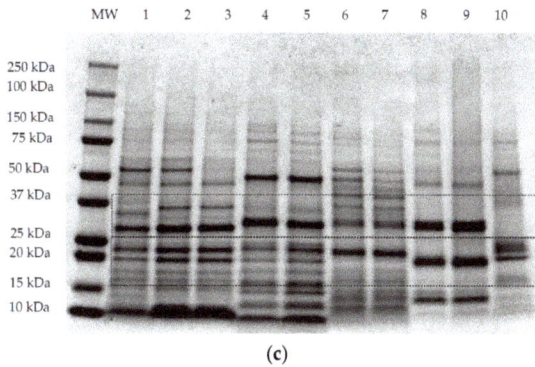

(c)

Figure 3. Representative sodium dodecyl sulphate–polyacrylamide gel electrophoresis (SDS-PAGE) patterns of quinoa wholegrain flour (1), quinoa dehulled flour (2), quinoa protein-rich flour (3), buckwheat dehulled flour (4), buckwheat protein-rich flour (5), amaranth wholegrain flour (6), amaranth protein-rich flour (7), rice flour (8), rice protein concentrate (9), and maize flour (10). The first lane of each gel contains the molecular weight marker. Samples were prepared according to methods I, II, and III for gel (**a–c**), respectively, as explained in Section 2.3.

When the samples were treated with a reducing protein extracting buffer (Figure 3c), the 50 kDa band was disrupted into several bands of lower MW and two of those bands were predominantly around 25–30 kDa and 15–20 kDa, corresponding to the subunits (α- or acidic and β- or basic) that form the globulins for pseudocereals or the glutelins for rice. For quinoa samples treated with the extracting buffer containing DTT as the reducing agent (Figure 3c), it was observed that the disulphide bonds that link the acidic or α- (MW ~28 and 34 kDa) and basic or β- (MW ~17 and 19 kDa) subunits were disrupted, leading to the dissociation of chenopodin into lower MW constituent proteins [28]. The same was observed for amaranth, whereby the acidic or α- (34–36 kDa) and basic or β- (22–24 kDa) subunits of amarantin linked by disulphide bonds are resolved under reducing conditions [17]. Buckwheat 13S legume-like globulin also consists of a small basic subunit (16–29 kDa) linked by a disulphide bond to a large acidic (30–38 kDa) subunit (Figure 3c) [52]. In the case of rice proteins, when the samples are treated with the reducing agent (Figure 3c), the glutelin precursor is disrupted into two main bands with MW ~30 and 20 kDa corresponding to the acidic (α-glutelin) and basic (β-glutelin) subunits that are linked by disulphide bonds. For maize proteins, when the sample was treated with the reducing extracting buffer (Figure 3c), two main protein bands were resolved around 20 kDa that may be related to the main maize protein, zein, a prolamin-like protein that accounts for 60% of the total protein [53].

Bands corresponding to low MW proteins (~10–15 kDa) could be observed in the three gels (Figure 3a–c) for all quinoa, amaranth, and buckwheat samples, which might be related to the albumin fraction, which is abundant in pseudocereals [54–56]. For rice samples the band evident at 13 kDa was reported previously as the prolamin fraction [20]. Besides globulin and albumin proteins, amaranth showed high MW proteins (~250 kDa; Figure 3a,b) which were resolved into bands of lower MW under reducing conditions (Figure 3c). Abugoch et al. [34], reported that amaranth glutelin contained an appreciable proportion of aggregated polypeptides of MW greater than 60 kDa. It is possible that the band evident on the gels at ~37 kDa for AWGF sample, and which is not disrupted under reducing conditions, might be the albumin-1 fraction, reported previously to have a MW of 34 kDa [17,55].

3.3. Starch Granules: Shape and Size

Different sizes, shapes, and structures were observed for flour and ingredient powder morphology and ultra-structure using scanning electron microscopy (SEM) analysis (Figure 4). Quinoa samples presented the smallest sized granules (1–1.20 μm) among all samples and had a polygonal shape.

The protein-rich flour (QPRF) showed granules covered and linked to other types of substances. This embryo-rich fraction is rich in protein, fibre, and fat which suggests that the starch granules are embedded in a matrix formed by these compounds. Li and Zhu [57] observed that some starch aggregates appeared to be coated with a film-like substance surrounded by a protein matrix. Amaranth samples, AWGF and APRF, showed circular granules with a size of ~2.5–3 µm. Amaranth seed is one of the few sources of small-granule starch, typically 1 to 3 µm in diameter, with a regular granule size [19]. The starch granules in APRF also appeared to be embedded within a matrix as observed for QPRF. Buckwheat starch granules showed the largest size (5 to 7.5 µm) among the pseudocereal samples with a mixture of spherical and polygonal structures. Christa et al. [58], also observed spherical, oval, and polygonal granules with a size distribution from 2 to 6 µm for buckwheat starch. Analysis of the granule structure and matrix positioning showed other components attached which may be protein and fat [59]. The BPRF samples, similar to that observed for QPRF and APRF, also had starch granules embedded in a matrix of other components. Analysis of RF ultrastructure showed starch granules with diameter between 4 and 5 µm, with an angular shape, while maize flour exhibited the largest starch granules (15 µm) with both circular and rod-shapes. These results are in agreement with Nienke et al. [60], who categorized starch granules into different sizes and defined the starch granules for amaranth and quinoa as very small, rice and buckwheat as small, and maize as generally having relatively large granules. The small size of the starch granules of some pseudocereals, such as quinoa, can offer advantages (e.g., altered emulsion stabilisation properties) in respect of incorporation into product formulations [57,61].

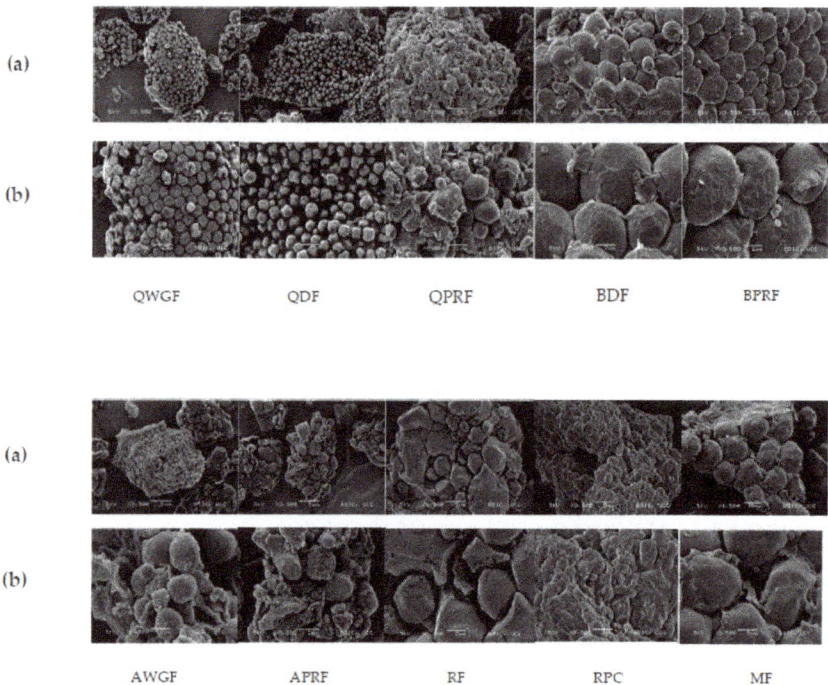

Figure 4. Scanning electron micrographs of quinoa wholegrain flour (QWGF), quinoa dehulled flour (QDF), quinoa protein-rich flour (QPRF), buckwheat dehulled flour (BDF), buckwheat protein-rich flour (BPRF), amaranth wholegrain flour (AWGF), amaranth protein-rich flour (APRF), rice flour (RF), rice protein concentrate (RPC), and maize flour (MF). *Magnification* row (**a**) ×3500; (**b**) ×8500. *Scale bars* row (**a**) 5 µm; (**b**) 2 µm.

3.4. Pasting Properties

The mean values for the initial, peak and final viscosity at the end of the holding stage at 95 °C, on completion of cooling to 50 °C, and at the end of the final holding period at 50 °C were recorded during pasting and are presented in Table 2. The shape of the pasting curves differed depending on the type of flour/ingredient (Figure 5a,b). Among the regular protein content flour samples, BDF and RF had the highest viscosity and AWGF the lowest. QWGF and QDF showed slight differences, with QDF having the lowest viscosity; this may be explained by the lower content of starch (50.5%) in QDF than in QWGF (60%). The peak time was very similar for all the flours (~12 min) tested, except MF and BDF which required a shorter time (~10.5 min) to reach peak viscosity, most likely due to the lower extent of absorption and swelling of their starch granules [62]. During the holding period at 95 °C, the material slurries were subjected to high temperature and mechanical shear stress, which further disrupted the starch granules, resulting in the leaching out from starch granules, and alignment, of amylose. It was observed that all the samples displayed a decrease in viscosity, especially so for BDF, RF, and MF, which had the more pronounced decreases in viscosity during the holding period at 95 °C (Figure 5a). The decrease in viscosity during the holding period is often correlated with high peak viscosity: it can be seen how BDF, RF and MF had the highest peak viscosities (Figure 5a). During cooling, re-association between starch molecules, especially amylose chains, will result in the formation of a gel structure and, therefore, viscosity will increase due to retrogradation and reordering of starch molecules. BDF (13.0 Pa·s) and AWGF (1.72 Pa·s) showed the highest and lowest final viscosity, respectively, while QWGF, QDF, RF, and MF showed broadly similar final viscosity values (5.83, 4.40, 4.31, and 5.07 Pa·s, respectively). Regarding the rheological profile of the protein-rich ingredients (Figure 5b), a similar pattern was observed as for the regular protein-content flours in respect of the initial, peak and final viscosities, but with considerably lower viscosity values observed overall. This can be explained by the lower content of starch and the higher content of dietary fibre in the protein-rich samples (Table 2). The water binding capacity of dietary fibre is greatly increased by the presence of high amounts of hydroxyl groups and can be related to a reduction in water availability, which could impact viscosity and pasting properties [63]. Also, the protein-rich flour ingredients are rich in ash, protein, and fat, which have been shown previously to influence the functionality of starch and impact on rheological behaviour of starch dispersions during pasting [59].

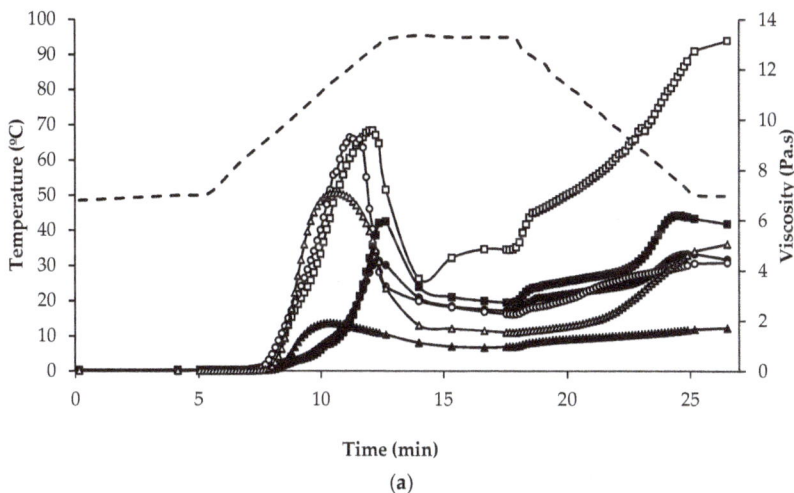

Time (min)

(a)

Figure 5. *Cont.*

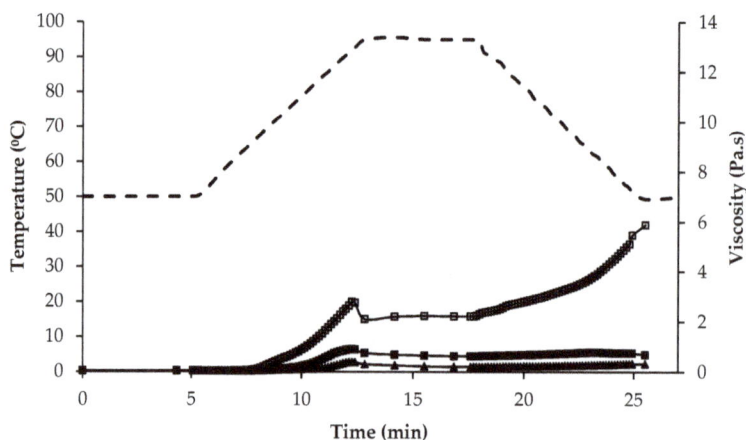

(b)

Figure 5. Temperature (dashed line) and viscosity (symbols) at various stages of the pasting regime of (a) regular protein containing flours: quinoa wholegrain flour (——■——), quinoa dehulled flour (——•——), amaranth wholegrain flour (——▲——), buckwheat dehulled flour (BDF) (——□——), rice flour (RF) (——○——), and maize flour (——△——); (b) of protein enriched flour ingredients: quinoa protein-rich flour (——•——), amaranth protein-rich flour (——▲——), buckwheat protein-rich flour (——□——).

Of particular interest, were the high and low viscosity values recorded during pasting for buckwheat and amaranth, respectively. These differences can be related to several factors associated with the starch component of the ingredients, namely the proportion and type of crystalline organization (amylose:amylopectin ratio), size and ultra-structure of the starch granule and extent of starch damage. The amylose content of amaranth and quinoa starch, a component which is related to a stronger and more cohesive gel with higher final paste viscosity [62,64], has been reported to be much lower than that found in buckwheat, rice, or maize [9]. In the case of quinoa starch, the amylose content ranges from 3.5% to 19.6% of total starch, while in amaranth seeds amylose levels have been reported to be lower than 8% [26]. In contrast, the amylose content of buckwheat has been reported to be as high as 57% [58]. Therefore, for buckwheat a higher final viscosity would be expected than for quinoa or amaranth. The starch granule size also influences the pasting temperature, whereby smaller granules have been associated with lower pasting temperatures [60]. BDF had the largest starch granules among the pseudocereal samples analysed in this study (Figure 4) while quinoa and amaranth had the smallest. Yoshimoto et al. [65], reported a higher granule swelling and gelling capacity for buckwheat starches compared with cereal starches. Another factor that can impact the pasting properties is the resistance of starch to digestion by α-amylase during the heating process; Izydorczyk et al. [66] associated the ability of buckwheat to form strong gels with the high resistance of the starch component to digestion by α-amylase. In addition, Lu et al. [67] associated reduced enzyme digestibility of cooked buckwheat groats with retrogradation and formation of resistant starch.

The understanding of the heat-induced rheological behaviour of these protein-rich ingredients is of great importance for the development of tailored nutritional products (e.g., low viscosity in plant-based milk substitutes or high viscosity in yogurt-type products).

Table 2. Viscosity of quinoa wholegrain flour (QWGF), quinoa dehulled flour (QDF), quinoa protein-rich flour (QPRF), amaranth wholegrain flour (AWGF), amaranth protein-rich flour (APRF), buckwheat dehulled flour (BDF), buckwheat protein-rich flour (BPRF), rice flour (RF), rice protein concentrate (RPC), and maize flour (MF) dispersions at various stages of the pasting regime. Values are means ± standard deviations of data from triplicate analysis.

	Initial Viscosity (mPa·s)	Peak Viscosity (Pa·s)	Peak Time (min)	Stage of Pasting		
				End of Holding at 95 °C (Pa·s)	End of Cooling to 50 °C (Pa·s)	Final Paste at 50 °C (Pa·s)
Quinoa						
QWGF	18.4 ± 0.76 [a,b]	6.25 ± 0.18 [e]	12.5	2.74 ± 0.34 [e]	6.07 ± 0.30 [e]	5.83 ± 0.43 [e]
QDF	24.8 ± 1.05 [b,c]	4.39 ± 0.16 [d]	12.5	2.41 ± 0.24 [c]	4.68 ± 0.19 [c]	4.40 ± 0.22 [c]
QPRF	24.1 ± 0.12 [b,c]	0.91 ± 0.03 [ab]	12.5	0.63 ± 0.01 [a]	0.75 ± 0.01 [a]	0.67 ± 0.01 [a]
Amaranth						
AWGF	29.1 ± 1.32 [c]	1.92 ± 0.07 [b,c]	10.3	0.98 ± 0.03 [b]	1.64 ± 0.06 [b]	1.72 ± 0.07 [b]
APRF	26.4 ± 0.84 [c]	0.29 ± 0.03 [a]	12.5	0.13 ± 0.01 [a]	0.19 ± 0.01 [a]	0.19 ± 0.01 [a]
Buckwheat						
BDF	48.4 ± 2.64 [e]	9.60 ± 0.46 [f]	12.1	4.84 ± 0.90 [f]	12.5 ± 0.47 [f]	13.0 ± 0.35 [f]
BPRF	37.5 ± 2.21 [d]	2.81 ± 0.10 [c]	12.5	2.23 ± 0.07 [d]	5.35 ± 0.24 [d]	5.92 ± 0.24 [e]
Rice						
RF	15.5 ± 0.06 [a]	9.37 ± 1.43 [f]	11.2	2.28 ± 0.23 [c]	4.24 ± 0.15 [c]	4.31 ± 0.13 [c]
RPC	17.9 ± 0.13 [a]	n.d.	n.d.	0.02 ± 0.00 [g]	0.02 ± 0.00 [g]	0.02 ± 0.00 [g]
Maize						
MF	16.4 ± 0.02 [a]	7.11 ± 0.20 [e]	10.6	1.54 ± 0.15 [cd]	4.67 ± 0.30 [cd]	5.07 ± 0.33 [d]

Values followed by different superscript letters (a–g) in the same column are significantly different ($p < 0.05$). n.d. = not detected.

4. Conclusions

In this study, the nutrient composition, protein profile, and rheological properties of a range of novel protein-rich pseudocereal flour ingredients were studied and compared to regular protein content pseudocereal, maize, and rice flours. The protein-rich flour ingredients had higher levels of ash, fat, and dietary fibre, and lower levels of starch. An integrated proteomic approach was implemented to gain enhanced clarity on the ingredient's protein profiles, with two strong protein extracting buffers being used for the first time, to allow the complete solubilization and characterization of the proteins in the pseudocereal ingredients. The results showed common bands under non-reducing and reducing conditions that corresponded to the globulin and albumin fractions. The predominance of globulins and albumins in pseudocereals is technologically significant since they are highly soluble in water and dilute salt solutions, which can be an advantage for food formulation purposes, in particular for the production of plant-based beverages. Buckwheat and amaranth had the highest and lowest final viscosity, respectively; while the protein-rich flours had considerably lower viscosity than their regular protein content counterparts. This study provides essential and much-needed new fundamental and applied knowledge on the compositional, structural, and functional properties of protein-rich pseudocereal ingredients to assist in further developing their utilisation in nutritious, functional, and stable food formulations.

Author Contributions: James A. O'Mahony and Loreto Alonso-Miravalles conceived and designed the experiments; Loreto Alonso-Miravalles performed the experiments, collated and analysed the data; James A. O'Mahony and Loreto Alonso-Miravalles prepared the manuscript.

Acknowledgments: This study was part of the PROTEIN2FOOD project. This project has received funding from the European Union's Horizon 2020 research and innovation programme under grant agreement No. 635727. The authors would like to acknowledge Juergen Bez (Fraunhofer Institut, Munich, Germany) for providing the protein-rich ingredients.

Conflicts of Interest: The authors declare no conflict of interest.

References

1. Henchion, M.; Hayes, M.; Mullen, A.M.; Fenelon, M.; Tiwari, B. Future protein supply and demand: strategies and factors influencing a sustainable equilibrium. *Foods* **2017**, *6*, 53. [CrossRef] [PubMed]
2. Day, L. Proteins from land plants—Potential resources for human nutrition and food security. *Trends Food Sci. Technol.* **2013**, *32*, 25–42. [CrossRef]
3. Nations, U. *Revision of World Population Prospects*; United Nations: New York, NY, USA, 2015.
4. Aiking, H. Future protein supply. *Trends Food Sci. Technol.* **2011**, *22*, 112–120. [CrossRef]
5. Schutyser, M.A.I.; Pelgrom, P.J.M.; van der Goot, A.J.; Boom, R.M. Dry fractionation for sustainable production of functional legume protein concentrates. *Trends Food Sci. Technol.* **2015**, *45*, 327–335. [CrossRef]
6. Wouters, A.G.B.; Rombouts, I.; Fierens, E.; Brijs, K.; Delcour, J.A. Relevance of the functional properties of enzymatic plant protein hydrolysates in food systems. *Compr. Rev. Food Sci. Food Saf.* **2016**, *15*, 786–800. [CrossRef]
7. Alting, A.C.; Van De Velde, F. Proteins as clean label ingredients in foods and beverages. In *Natural Food Additives, Ingredients and Flavourings*; Baines, D., Seal, R., Eds.; Woodhead Publishing Limited: Cambridge, UK, 2012; pp. 197–211. ISBN 978-1-84-569811-9.
8. Jacobsen, S.E.; Sørensen, M.; Pedersen, S.M.; Weiner, J. Feeding the world: Genetically modified crops versus agricultural biodiversity. *Agron. Sustain. Dev.* **2013**, *33*, 651–662. [CrossRef]
9. Haros, C.M.; Schoenlechner, R. *Pseudocereals: Chemistry and Technology*, 1st ed.; Wiley Blackwell: Chichester, UK, 2017; ISBN 978-1-11-893825-6.
10. Schoenlechner, R.; Siebenhandl, S.; Berghofer, E. Pseudocereals. In *Gluten-Free Cereal Product and Beverages*; Arendt, E.K., Dal Bello, F., Eds.; Elsevier: New York, NY, USA, 2008; pp. 149–190. ISBN 978-0-12-373739-7.
11. Taylor, J.; Awika, J. *Gluten-Free Ancient Grains. Cereals, Pseudocereals, and Legumes: Sustainable, Nutritious, and Health-Promoting Foods for the 21st Century*, 1st ed.; Taylor, J., Awika, J., Eds.; Woodhead Publishing Limited: Duxford, UK, 2017; ISBN 978-0-08-100866-9.

12. Mota, C.; Santos, M.; Mauro, R.; Samman, N.; Matos, A.S.; Torres, D.; Castanheira, I. Protein content and amino acids profile of pseudocereals. *Food Chem.* **2016**, *193*, 55–61. [CrossRef] [PubMed]

13. Tester, R.F.; Karkalas, J.; Qi, X. Starch—Composition, fine structure and architecture. *J. Cereal Sci.* **2004**, *39*, 151–165. [CrossRef]

14. Steadman, K.J.J.; Burgoon, M.S.S.; Lewis, B.A.A.; Edwardson, S.E.E.; Obendorf, R.L.L. Buckwheat seed milling fractions: Description, macronutrient composition and dietary fibre. *J. Cereal Sci.* **2001**, *33*, 271–278. [CrossRef]

15. Schirmer, M.; Höchstötter, A.; Jekle, M.; Arendt, E.; Becker, T. Physicochemical and morphological characterization of different starches with variable amylose/amylopectin ratio. *Food Hydrocoll.* **2013**, *32*, 52–63. [CrossRef]

16. Horstmann, S.; Belz, M.M.; Heitmann, M.; Zannini, E.; Arendt, E. Fundamental study on the impact of gluten-free starches on the quality of gluten-free model breads. *Foods* **2016**, *5*, 30. [CrossRef] [PubMed]

17. Janssen, F.; Pauly, A.; Rombouts, I.; Jansens, K.J.A.; Deleu, L.J.; Delcour, J.A. Proteins of amaranth (*Amaranthus* spp.), buckwheat (*Fagopyrum* spp.), and quinoa (*Chenopodium* spp.): A food science and technology perspective. *Compr. Rev. Food Sci. Food Saf.* **2017**, *16*, 39–58. [CrossRef]

18. Mäkinen, O.E.; Zannini, E.; Koehler, P.; Arendt, E.K. Heat-denaturation and aggregation of quinoa (*Chenopodium quinoa*) globulins as affected by the pH value. *Food Chem.* **2016**, *196*, 17–24. [CrossRef] [PubMed]

19. Venskutonis, P.R.; Kraujalis, P. Nutritional components of amaranth seeds and vegetables: A review on composition, properties, and uses. *Compr. Rev. Food Sci. Food Saf.* **2013**, *12*, 381–412. [CrossRef]

20. Amagliani, L.; O'Regan, J.; Kelly, A.L.; O'Mahony, J.A. Composition and protein profile analysis of rice protein ingredients. *J. Food Compos. Anal.* **2016**, *59*, 18–26. [CrossRef]

21. DeVries, J.W. *Dietary Fibre: New Frontiers for Food and Health*; van der Kamp, J.W., Jones, J.M., McCleary, B.V., Topping, D.L., Eds.; Wageningen Academic Publishers: Wageningen, The Netherlands, 2010; ISBN 978-9-08-686128-6.

22. Codex Alimentarius Commission. *Report of the 27th Session of the Codex Committee on Nutrition and Foods for Special Dietary Uses*; Codex Alimentarius Commission: Rome, Italy, 2005.

23. Foschia, M.; Peressini, D.; Sensidoni, A.; Brennan, C.S. The effects of dietary fibre addition on the quality of common cereal products. *J. Cereal Sci.* **2013**, *58*, 216–227. [CrossRef]

24. Alvarez-Jubete, L.; Arendt, E.K.; Gallagher, E. Nutritive value and chemical composition of pseudocereals as gluten-free ingredients. *Int. J. Food Sci. Nutr.* **2009**, *60*, 240–257. [CrossRef] [PubMed]

25. Burrieza, H.P.; Lopez-Fernandez, M.P.; Maldonado, S. Analogous reserve distribution and tissue characteristics in quinoa and grass seeds suggest convergent evolution. *Front. Plant Sci.* **2014**, *5*, 546. [CrossRef] [PubMed]

26. Alvarez-Jubete, L.; Arendt, E.K.; Gallagher, E. Nutritive value of pseudocereals and their increasing use as functional gluten-free ingredients. *Trends Food Sci. Technol.* **2010**, *21*, 106–113. [CrossRef]

27. Nosworthy, M.G.; Tulbek, M.C.; House, J.D. Does the concentration, isolation, or deflavoring of pea, lentil, and faba bean protein alter protein quality? *Cereal Foods World* **2017**, *62*, 139–142. [CrossRef]

28. Avila Ruiz, G.; Arts, A.; Minor, M.; Schutyser, M. A hybrid dry and aqueous fractionation method to obtain protein-rich fractions from quinoa (*Chenopodium quinoa* Willd). *Food Bioprocess Technol.* **2016**, *9*, 1502–1510. [CrossRef]

29. Berghout, J.A.M.; Pelgrom, P.J.M.; Schutyser, M.A.I.; Boom, R.M.; van der Goot, A.J. Sustainability assessment of oilseed fractionation processes: A case study on lupin seeds. *J. Food Eng.* **2015**, *150*, 117–124. [CrossRef]

30. Boukid, F.; Folloni, S.; Sforza, S.; Vittadini, E.; Prandi, B. Current trends in ancient grains-based foodstuffs: insights into nutritional aspects and technological applications. *Compr. Rev. Food Sci. Food Saf.* **2018**, *17*, 123–136. [CrossRef]

31. Pelgrom, P.J.M.; Boom, R.M.; Schutyser, M.A.I. Functional analysis of mildly refined fractions from yellow pea. *Food Hydrocoll.* **2015**, *44*, 12–22. [CrossRef]

32. AOAC. *Official Methods of Analysis of the Association of Official Analytical Chemists*, 18th ed.; AOAC: Washington, DC, USA, 2010.

33. Nascimento, A.C.; Mota, C.; Coelho, I.; Gueifão, S.; Santos, M.; Matos, A.S.; Gimenez, A.; Lobo, M.; Samman, N.; Castanheira, I. Characterisation of nutrient profile of quinoa (*Chenopodium quinoa*), amaranth (*Amaranthus caudatus*), and purple corn (*Zea mays* L.) consumed in the North of Argentina: Proximates, minerals and trace elements. *Food Chem.* **2014**, *148*, 420–426. [CrossRef] [PubMed]

34. Abugoch James, L.E. Quinoa (*Chenopodium quinoa* Willd.): Composition, chemistry, nutritional, and functional properties. *Adv. Food Nutr. Res.* **2009**, *58*, 1–31. [CrossRef] [PubMed]
35. Li, G.; Wang, S.; Zhu, F. Physicochemical properties of quinoa starch. *Carbohydr. Polym.* **2016**, *137*, 328–338. [CrossRef] [PubMed]
36. Bradley, R.L. Moisture and Total Solids Analysis. In *Food Analysis*; Nielsen, S., Ed.; Springer: West Lafayette, IN, USA, 2010; pp. 85–104. ISBN 1441914781.
37. Roudaut, G.; Debeaufort, F. Moisture loss, gain and migration in foods. In *Food and Beverage Stability and Shelf Life*; Elsevier: New York, NY, USA, 2011; pp. 63–105. ISBN 978-1-84-569701-3.
38. Schuck, P.; Dolivet, A.; Jeantet, R. Determination of Dry Matter and Total Dry Matter. In *Analytical Methods for Food and Dairy Powders*; Schuck, P., Dolivet, A., Jeantet, R., Eds.; John Wiley & Sons, Ltd.: Hoboken, NJ, USA, 2012; pp. 45–57. ISBN 978-1-11-830739-7.
39. Tyler, R.T.; Youngs, C.G.; Sosulski, F.W. Air Classification of legumes. I. Separation efficiency, yield, and composition of the starch and protein fractions. *Cereal Chem.* **1981**, *58*, 144–148.
40. Arendt, E.K.; Zannini, E. *Cereal Grains for the Food and Beverage Industries*, 1st ed.; Woodhead Publishing Limited: Cambridge, UK, 2013; ISBN 978-0-85-709413-1.
41. Gamel, T.H.; Mesallam, A.S.; Damir, A.A.; Shekib, L.A.; Linssen, J.P. Characterization of amaranth seeds oil. *J. Food Lipids* **2007**, *14*, 323–334. [CrossRef]
42. Navruz-Varli, S.; Sanlier, N. Nutritional and health benefits of quinoa (*Chenopodium quinoa* Willd.). *J. Cereal Sci.* **2016**, *69*, 371–376. [CrossRef]
43. Barrera, G.N.; Bustos, M.C.; Iturriaga, L.; Flores, S.K.; Leon, A.E.; Ribotta, P.D. Effect of damaged starch on the rheological properties of wheat starch suspensions. *J. Food Eng.* **2013**, *116*, 233–239. [CrossRef]
44. Nowak, V.; Du, J.; Charrondière, U.R. Assessment of the nutritional composition of quinoa (*Chenopodium quinoa* Willd.). *Food Chem.* **2016**, *193*, 47–54. [CrossRef] [PubMed]
45. Ando, H.; Chen, Y.-C.; Tang, H.; Mayumi, S.; Watanabe, K.; Mitsunaga, T. Food components in fractions of quinoa seed. *Food Sci. Technol. Res.* **2002**, *8*, 80–84. [CrossRef]
46. Repo-Carrasco, R.; Peña, J.; Kallio, H.; Salminen, S. Dietary fiber and other functional components in two varieties of crude and extruded kiwicha (*Amaranthus caudatus*). *J. Cereal Sci.* **2009**, *49*, 219–224. [CrossRef]
47. Ruales, J.; Grijalva, Y.; Lopez-Jaramillo, P.; Nair, B. The nutritional quality of an infant food from quinoa and its effect on the plasma level of insulin-like growth factor-1 (IGF-1) in undernourished children. *Int. J. Food Sci. Nutr.* **2002**, *53*, 143–154. [CrossRef] [PubMed]
48. Repo-Carrasco, R.; Astuhuaman Serna, L. Quinoa (*Chenopodium quinoa*, Willd.) as a source of dietary fiber and other functional components. *Food Sci. Technol.* **2009**, *31*, 225–230. [CrossRef]
49. Mir, N.A.; Riar, C.S.; Singh, S. Nutritional constituents of pseudo cereals and their potential use in food systems: A review. *Trends Food Sci. Technol.* **2018**, *75*, 170–180. [CrossRef]
50. Vasco-Méndez, N.L.; Paredes-López, O. Antigenic homology between amaranth glutelins and other storage proteins. *Food Biochem.* **1994**, *18*, 227–238. [CrossRef]
51. Milisavljević, M.D.; Timotijević, G.S.; Radović, S.R.; Brkljačić, J.M.; Konstantinović, M.M.; Maksimović, V.R. Vicilin-like storage globulin from buckwheat (*Fagopyrum esculentum* Moench) seeds. *J. Agric. Food Chem.* **2004**, *52*, 5258–5262. [CrossRef] [PubMed]
52. Choi, S.M.; Ma, C.Y. Conformational study of globulin from common buckwheat (*Fagopyrum esculentum* Moench) by fourier transform infrared spectroscopy and differential scanning calorimetry. *J. Agric. Food Chem.* **2005**, *53*, 8046–8053. [CrossRef] [PubMed]
53. Anderson, T.J.; Lamsa, B.P. Zein extraction from corn, corn products, and coproducts and modifications for various applications: A review. *Cereal Chem.* **2011**, *88*, 159–173. [CrossRef]
54. Valencia-Chamorro, S.A. Quinoa. In *Encyclopedia of Food Sciences and Nutrition*; Elsevier: Amsterdam, The Netherlands, 2003; pp. 4895–4902.
55. Gorinstein, S.; Pawelzik, E.; Delgado-Licon, E.; Haruenkit, R.; Weisz, M.; Trakhtenberg, S. Characterisation of pseudocereal and cereal proteins by protein and amino acid analyses. *J. Sci. Food Agric.* **2002**, *82*, 886–891. [CrossRef]
56. Radovic, R.S.; Maksimovic, R.V.; Brkljacic, M.J.; Varkonji Gasic, I.E.; Savic, P.A. 2S albumin from buckwheat (*Fagopyrum esculentum* Moench) seeds. *J. Agric. Food Chem.* **1999**, *47*, 1467–1470. [CrossRef] [PubMed]
57. Li, G.; Zhu, F. Molecular structure of quinoa starch. *Carbohydr. Polym.* **2017**, *158*, 124–132. [CrossRef] [PubMed]

58. Christa, K.; Soral-Smietana, M.; Lewandowicz, G. Buckwheat starch: Structure, functionality and enzyme in vitro susceptibility upon the roasting process. *Int. J. Food Sci. Nutr.* **2009**, *60*, 140–154. [CrossRef] [PubMed]

59. Debet, M.R.; Gidley, M.J. Three classes of starch granule swelling: Influence of surface proteins and lipids. *Carbohydr. Polym.* **2006**, *64*, 452–465. [CrossRef]

60. Nienke, L.; Chang, P.R.; Tyler, R.T. Analytical, biochemical and physicochemical aspects of starch granule size, with emphasis on small granule starches: A review. *Starch Stärke* **2004**, *56*, 89–99. [CrossRef]

61. Rayner, M.; Timgren, A.; Sjoo, M.; Dejmek, P. Quinoa starch granules: A candidate for stabilising food-grade pickering emulsions. *J. Sci. Food Agric.* **2012**, *92*, 1841–1847. [CrossRef] [PubMed]

62. Ragaee, S.; Abdel-Aal, E.S.M. Pasting properties of starch and protein in selected cereals and quality of their food products. *Food Chem.* **2006**, *95*, 9–18. [CrossRef]

63. Bulut-Solak, B.; Alonso-Miravalles, L.; O'Mahony, J.A. Composition, morphology and pasting properties of Orchis anatolica tuber gum. *Food Hydrocoll.* **2016**, *69*, 483–490. [CrossRef]

64. Wang, S.; Li, C.; Copeland, L.; Niu, Q.; Shuo, W. Starch retrogradation: A comprehensive review. *Compr. Rev. Food Sci. Food Saf.* **2015**, *14*, 568–585. [CrossRef]

65. Yoshimoto, Y.; Egashira, T.; Hanashiro, I.; Ohinata, H.; Takase, Y.; Takeda, Y. Molecular structure and some physicochemical properties of buckwheat starches. *Cereal Chem.* **2004**, *81*, 515–520. [CrossRef]

66. Izydorczyk, M.S.; McMillan, T.; Bazin, S.; Kletke, J.; Dushnicky, L.; Dexter, J. Canadian buckwheat: A unique, useful and under-utilized crop. *Can. J. Plant Sci.* **2013**, *94*, 509–524. [CrossRef]

67. Lu, L.; Murphy, K.; Baik, B.K. Genotypic variation in nutritional composition of buckwheat groats and husks. *Cereal Chem.* **2013**, *90*, 132–137. [CrossRef]

![foods logo] *foods*

MDPI

Review

Use of Sourdough in Low FODMAP Baking

Jussi Loponen [1] and Michael G. Gänzle [2,*]

[1]   Fazer Group, 01230 Vantaa, Finland; jussi.loponen@fazer.com
[2]   Department of Agricultural, Food and Nutritional Science, University of Alberta,
       Edmonton, AB T6G 2P5, Canada
*   Correspondence: mgaenzle@ualberta.ca; Tel.: +1-780-492-0774

Received: 3 June 2018; Accepted: 19 June 2018; Published: 22 June 2018

Abstract: A low FODMAP (fermentable oligosaccharides, disaccharides, monosaccharides, and polyols) diet allows most irritable bowel syndrome (IBS) patients to manage their gastrointestinal symptoms by avoiding FODMAP-containing foods, such as onions, pulses, and products made from wheat or rye. The downside of a low FODMAP diet is the reduced intake of dietary fiber. Applying sourdoughs—with specific FODMAP-targeting metabolic properties—to wholegrain bread making can help to remarkably reduce the content of FODMAPs in bread without affecting the content of the slowly fermented and well-tolerated dietary fiber. In this review, we outline the metabolism of FODMAPs in conventional sourdoughs and outline concepts related to fructan and mannitol metabolism that allow development of low FODMAP sourdough bread. We also summarize clinical studies where low FODMAP but high fiber, rye sourdough bread was tested for its effects on gut fermentation and gastrointestinal symptoms with very promising results. The sourdough bread-making process offers a means to develop natural and fiber-rich low FODMAP bakery products for IBS patients and thereby help them to increase their dietary fiber intake.

Keywords: sourdough; FODMAP; fructan; mannitol; lactobacilli; irritable bowel syndrome (IBS); non-celiac wheat intolerance

1. Introduction

Fermentable oligosaccharides, disaccharides, monosaccharides, and polyols (FODMAPs) have beneficial and adverse health effects [1]. Oligosaccharides that are not hydrolyzed and absorbed in the small intestine are rapidly fermented by intestinal microbiota in the terminal ileum and the proximal colon [2,3]. Diverse FODMAPs that are fermented by intestinal microbiota consistently cause adverse symptoms when a dose of about 0.3 g/kg body weight, corresponding to about 15 g/day, is exceeded [4,5]. Adverse symptoms include osmotic diarrhea, intestinal distension, and bloating [5,6]. The extent of the adverse symptoms decreases with the degree of polymerization because of the reduced osmotic load of oligosaccharides in the small intestine, and the reduced rate of fermentation [6]. Adverse effects are not described for non-digestible polysaccharides, which are fermented at a much lower rate [7]. Microbiota in the terminal ileum include proteobacteria and lactic acid bacteria as the dominant representatives; ileal microbiota effectively ferment mono- and disaccharides but typically lack extracellular enzymes for hydrolysis of higher oligosaccharides and polysaccharides [6]. The sensitivity of individuals to adverse symptoms caused by FODMAPs is highly variable; adverse symptoms are often linked to irritable bowel syndrome (IBS). The sensitivity to gas pressure and pain varies highly among individuals; moreover, intestinal microbiota adapt toward the fermentation of specific oligosaccharides; this adaptation reduces or eliminates adverse symptoms [8]. Many FODMAPs are conditionally digestible depending on the genetic status of the host. About 35% of humans are lactase-persistent and digest lactose while lactose is a non-digestible FODMAP in the remainder of the population [9]. A substantial proportion of humans are fructose intolerant; the proportion of fructose intolerant individuals among patients with IBS was reported to be over 60% [10,11]. Fructose absorption is highly dependent on the presence of equimolar amounts of glucose as uptake from the small intestine uses the same transport channels [10]. A rare variation in the sucrose-isomaltase gene reduces the digestibility of sucrose, including sucrose in the FODMAPs; this genetic variant also predisposes for IBS [12].

Health beneficial or prebiotic effects of oligosaccharides relate to the bacterial conversion of oligosaccharides to short chain fatty acids [1,13]. These short chain fatty acids increase the energy harvest from carbohydrates that escape small intestinal hydrolysis and absorption, improve intestinal barrier properties and resistance to enteric infections, and exert systemic effects related to inflammation, cognitive functions, and behavior through specific recognition with G-protein coupled receptors (for reviews, see [1,7,13]). Of note, oligomeric fructans, for which health beneficial prebiotic effects were most consistently demonstrated [13], appear also of particular concern for adverse effects in IBS [6]. Adverse and beneficial effects of FODMAPs are thus interconnected and partially related to the same mechanisms, bacterial fermentation. Consequently, a reduction of adverse symptoms in IBS by a low FODMAP diet also increased the luminal pH and reduced the abundance of bifidobacteria and butyrate-producing colonic bacteria [14,15]. While the term FODMAPs indiscriminately includes all oligosaccharides, different compounds were reported to have divergent effects. Supplementation of a low FODMAP diet with β-galacto-oligosaccharides was reported to improve IBS symptoms relative to a low FODMAP diet [16]. In other words, replacement of FODMAPs with different categories of FODMAPs may improve symptoms of IBS without the adverse consequences of a low fiber diet [1].

Wheat and rye are major contributors to the dietary intake of low molecular weight fructans [17] but whole grain products also are major contributors to the intake of dietary fiber [7]. Fermentation processes during baking may allow conversion or degradation of FODMAPs without reducing the overall dietary fiber content of bread [18]. This review aims to summarize current knowledge on the use of conventional and sourdough baking in the production of low FODMAP bread.

2. FODMAPs as Contributors to Non-Celiac Wheat Sensitivity?

Non-celiac wheat sensitivity refers to syndromes where components of wheat cause intestinal symptoms. Triggers and mechanisms of the syndrome are poorly described; non-celiac wheat sensitivity is often self-diagnosed or assessed after exclusion of celiac disease and wheat allergy [19,20]. Non-celiac wheat sensitivity overlaps significantly with IBS [20]. Non-celiac wheat sensitivity has also been described as non-gluten wheat sensitivity since gluten apparently is not a major trigger in these symptoms [21]. While a contribution of FODMAPs to symptoms in IBS is increasingly supported by clinical trials, their role in non-celiac wheat sensitivity is not as well documented. FODMAPs and amylase trypsin inhibitors (ATIs) were suggested as likely non-gluten triggers of these symptoms [19,20]. It is likely that *Triticeae* cereals other than wheat, such as rye and barley, are also potential triggers of wheat sensitivity because they also contain fructans and ATIs.

3. FODMAPs in Cereals and FODMAP Metabolism in Conventional Sourdoughs

Resting grains of wheat and rye contain only low levels of monosaccharides; the major oligosaccharides are sucrose, raffinose, and fructans (Table 1). During sourdough fermentation, amylase and glucoamylase activities of wheat and rye flour release maltose and glucose, respectively, from damaged starch [18]. The fructans of cereal grains are graminan-type fructans, which are oligosaccharides built of mixed-linkage fructose units [22]. Fructans in wheat and rye are concentrated in the outer layers of the grain and have an average degree of polymerization (DP) of 5–6; 1-kestose and nystose account for only a minor proportion of the overall fructans (Table 1) [23]. Additional non-starch polysaccharides include arabinoxylans and β-glucans as the major components, polysaccharides composed of mannose, galactose, and galacturonic acid, and trace amounts of pectin (Table 1). In addition to polysaccharides and FODMAPs that are present in the grain, polysaccharides, oligosaccharides, and polyols can be produced by bacterial activity during sourdough fermentation.

An overview of the conversion and production of FODMAPs in sourdough fermentation is provided in Figure 1.

During bread making, the fructans undergo partial degradation due to invertase activity present in yeast. The remaining fructan has a lower DP than the native fructan of flour. Low molecular weight fructans may be under-estimated when analyzing fructan in dough; in addition, they are fermented more rapidly than fructans with a higher molecular weight. The fate of fructans is valid for sourdough fermentation, i.e., grain fructans degrade to some extent but in the case of sourdough, the released fructose is also partially converted to mannitol by sourdough lactobacilli. Mannitol is a polyol that is rapidly fermented by gut microbiota. Thus, for accurate FODMAP quantification, mannitol levels in sourdough breads should also be determined. In the following sections, we outline the carbohydrate metabolism in sourdoughs. This is relevant to understand when the focus is in changes of FODMAPs in sourdough bread making.

Table 1. Content of oligosaccharides and non-starch polysaccharides (%) in wheat and rye grains.

Saccharide	Wheat	Rye
Arabinoxylans	6–7	7–12
β-Glucans including lignified cellulose	0.3–3	2–3
Pectin	trace	trace
Mannans, galactans, and galacturonans	1–1.5	n.d.
Fructans	1–2	4.3–5
1-Kestose	0.1	0.3
Nystose	0.03	0.1
Sucrose	0.6–1.0	1.2–1.8
Maltose	trace	trace
Raffinose	0.2–0.7	0.1–0.7
Stachyose	trace	trace

Compiled with information from [17,23–31]; n.d., not determined.

In straight dough processes, the dough is fermented with baker's yeast as the sole fermentation organism; the addition of high cell counts of *S. cerevisiae*, 1–2% biomass corresponding to about 10^8 cfu/g, achieves leavening after a fermentation time of 2 h or less. In sourdough baking, lactic acid bacteria are used as the second group of organisms; moreover, part of the flour is fermented for an extended period of time. The inclusion of lactic acid bacteria extends the metabolic capacity of the fermentation microbiota; the extended fermentation time strongly enhances the contribution of flour enzymes to the conversion and degradation of dough components [18]. Type I sourdoughs are typically fermented between 15 and 30 °C and they have traditionally been used as the sole leavening agent in bread making. To ensure a sufficient metabolic activity and leavening capacity, type I sourdoughs are propagated through one to three fermentation steps prior to mixing the bread dough [27,32]. Fermentation procedures that use sourdough as the sole leavening agent typically result in ~10% of the flour being fermented for >12 h, 20–30% fermented for >6 h, and all of the flour fermented for 2–3 h, i.e., the time required for dough rest and proofing [33,34]. Fermentation organisms in type I sourdoughs generally include *Lactobacillus sanfranciscensis* and *Kazachstania humilis* (syn. *Candida milleri*) and *S. cerevisiae* or *S. exiguus*. Lactobacilli of the *L. brevis*, *L. plantarum*, and *L. reuteri* groups are also represented in type I sourdoughs [32,35]. Industrial bread production generally includes baker's yeast as the leavening agent; sourdough fermentations in industrial baking (type I or II sourdoughs) aim at dough acidification to improve the baking quality of rye flour, at supporting the leavening capacity of baker's yeast, and as baking improver [32–34]. Fermentation conditions depend on the technological aim of the fermentation and are often specific for a specific production site; typically, 5–20% of the flour is fermented for >12 h while the remainder of the flour is fermented for ~2 h, corresponding to dough rest, shaping, and proofing [33,34]. Type II sourdough fermentation takes place at around 40 °C and the microbiota typically comprise organisms of the *L. delbrueckii* group (e.g., *L. amylovorans* and *L. johnsonii*) and organisms of the *L. reuteri* group (e.g., *L. reuteri*, *L. pontis*, and *L. panis*) [32,36]. Sourdough microbiota are metabolically active if the sourdough is fermented at the bakery but inactivated if the sourdough is stabilized by drying or pasteurization prior to use in baking [34].

Sucrose is metabolized rapidly by invertase activity of *S. cerevisiae*. Yeast invertase is an extracellular or cell wall-bound enzyme and is secreted in excess of the yeast's capacity to ferment the hydrolysis products [37]. Sucrose metabolism in lactic acid bacteria is mediated by sucrose phosphorylase or sucrose-1-phosphate hydrolase [38]. Sucrose metabolism and the metabolism of other oligosaccharides in homofermentative lactic acid bacteria is repressed by glucose [39]; in contrast, sucrose conversion in heterofermentative lactic acid bacteria is induced by the substrate but not repressed by glucose [40,41]. Fructose is utilized as a carbon source by homofermentative lactic acid bacteria but used as an electron acceptor for the regeneration of reduced cofactors by most heterofermentative lactobacilli [41,42]. Sourdough lactic acid bacteria also harbor extracellular glucansucrases or fructansucrases, which convert sucrose to indigestible poly- and oligosaccharides. These enzymes are frequently present in *Leuconostoc* spp., *Weissella* spp., and species of the *L. reuteri* and *L. delbrueckii* groups but are also present in other lactobacilli including *L. sanfranciscensis* [43,44]. Glucansucrases convert sucrose to polymeric glucans, isomalto-oligosaccharides, and fructose; fructansucrases catalyze the conversion to levan or inulin, fructo-oligosaccharides, and glucose [44]. Sucrose conversion by glucansucrases and fructansucrases accumulated isomalto-oligosaccharides and fructo-oligosaccharides, respectively, in wheat and sorghum sourdoughs; however, accumulation of oligosaccharides to relevant concentrations is observed only when sucrose is added to the sourdough [45,46]. Glucansucrases and the hydrolase activity of fructansucrases generally also release fructose, which is converted to the polyol mannitol by heterofermentative lactic acid bacteria [41,43]. In traditional sourdough fermentations, mannitol accumulates to 10–20 mmol/kg in wheat and 50 mmol/kg in rye, corresponding to 0.2–0.4% and 0.9%, respectively; the mannitol concentration is increased in direct proportion to the sucrose addition to sourdoughs [47]. *Weissella* spp. are exceptional because the majority of strains do not produce mannitol from fructose [45].

Lactic acid bacteria metabolize raffinose by sequential activity of extracellular levansucrase to convert raffinose to melibiose and fructose or fructan, followed by melibiose transport and intracellular hydrolysis by α-galactosidase. An alternative pathway involves raffinose transport and sequential hydrolysis by intracellular α-galactosidase to convert raffinose to sucrose and galactose and sucrose phosphorylase [48]. Metabolism by extracellular levansucrase with intracellular α-galactosidase is faster than the alternate pathway using two intracellular enzymes, presumably because the disaccharide melibiose is transported faster than raffinose [48]. Raffinose metabolism in heterofermentative lactobacilli is not subject to carbon catabolite repression [49] and the relatively high concentrations of raffinose and raffinose level oligosaccharides in pulse flours are rapidly degraded during fermentation [48]. Type I sourdough microbiota and most strains of *S. cerevisiae* are raffinose negative. Nevertheless, levansucrase from *L. sanfranciscensis* and/or yeast invertase converts raffinose to fructose and melibiose [43,50].

The content of fructans is reduced in straight dough processing to 1–1.5% fructans in wheat bread and about 3% in rye bread [51]. Fructans are not degraded in simulated sourdoughs without microbial activity but invertase activity of *S. cerevisiae* and *Kazachstania humilis* results in partial hydrolysis of flour fructans [52,53]. In a straight dough process, the rate of fructan hydrolysis decreases in the order trisaccharides > tetrasaccharides > pentasaccharides and only a small proportion of higher fructans are degraded [54]. Hydrolysis of fructans is mediated by yeast. However, dimerization of the enzyme reduces the activity towards kestose and nystose and sterically prevents access of oligosaccharides with a DP of more than four to the catalytic site [55]. Metabolism of fructans in lactobacilli is mediated by oligosaccharide transport through the ATP-Bbinding-Cassette transporter MsmEFGK or the phosphotransferase (PTS) system PTS1Bca, followed by hydrolysis through intracellular fructosidases or phospho-fructosidases, respectively [38]. Oligosaccharide transport by MsmEFGK and PTS1BCA is limited to fructans with a DP of four or less [56,57]. Metabolic enzymes for fructo-oligosaccharide (FOS) catabolism are frequent in homofermentative lactobacilli where FOS degradation is repressed by glucose [58] but are very infrequently found in heterofermentative lactobacilli [38,43,49]. Intracellular metabolism of FOS by lactobacilli thus does not contribute to the degradation of fructans in wheat or rye sourdoughs.

In summary, conventional dough fermentations, including sourdough fermentations, result in decreased levels of FODMAPs but may generate FODMAPs from the digestible carbohydrates sucrose and fructose (Figure 1). Low FODMAP baking thus necessitates dedicated approaches, particularly involving fructan- and mannitol-degrading organisms.

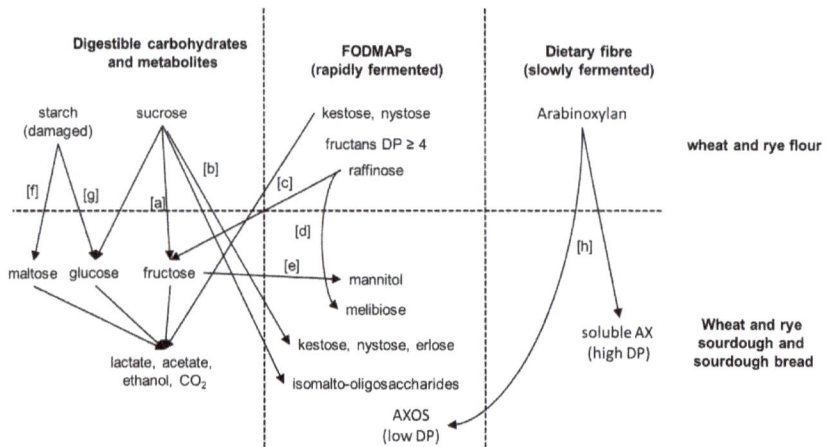

Figure 1. Conversion and generation of fermentable oligosaccharides, disaccharides, monosaccharides, and polyols (FODMAPs) in wheat and rye sourdoughs. Sucrose hydrolysis by yeast invertase or fructosidases of lactic acid bacteria [a]. Oligosaccharide formation by glucansucrases to form isomalto-oligosaccharides, or by fructansucrases to form kestose, nystose, and erlose from sucrose [b]. Kestose and nystose degradation by yeast invertase or by intracellular (phospho)-fructosidases of lactic acid bacteria [c]. Raffinose conversion by yeast invertase and levansucrase from lactic acid bacteria [d]. Fructose conversion by mannitol-dehydrogenase from heterofermentative lactic acid bacteria [e]. Starch conversion to maltose and glucose by flour amylases and gluco-amylase [f,g]. Exogenous xylanases are used in baking to increase the amount of soluble pentosane (arabinoxylan, AX) to improve bread properties, which can produce low DP arabinoxylan oligosaccharides (AXOS) along soluble high-DP arabinoxylan fragments [h].

4. Concepts for Low FODMAP Sourdough Baking

Degradation of fructans with a DP of more than four requires extracellular fructanases. Baker's yeast *S. cerevisiae* does not express extracellular fructanase. However, *Kluyveromyces marxianus* was suggested as an alternative leavening agent with extracellular fructanase activity [53,59]. *K. marxianus* is maltose negative and most strains do not provide sufficient CO_2 production for dough leavening; the use of *K. marxianus* in low FODMAP baking thus requires co-culture with *S. cerevisiae* [53] or selection of *K. marxianus* strains with sufficient leavening power and addition of amyloglucosidase to provide glucose for *K. marxianus* metabolism [53,59]. Dough fermentation with *K. marxianus* alone or in co-culture with *S. cerevisiae* allowed production of experimental breads with a low fructan content and a volume and sensory properties matching those of experimental breads produced with baker's yeast [53,59].

Extracellular glycosyl hydrolases are exceptional in lactobacilli [38]; accordingly, only a few strains with extracellular fructanase activity have been characterized (Figure 2). The extracellular

GH32 β-fructanase FosE was characterized in *L. paracasei* [60]. FosE is an extracellular enzyme that is induced by fructose, sucrose, or inulin but repressed by glucose [60]. BLAST analysis frequently identified homologues of this enzyme in other strains of the *L. casei* group and in few strains of the *L. salivarius* group (Figure 2 and data not shown). The β-fructanase FruA of *Streptococcus mutans* is extracellular with an LPXTG cell wall anchor; the enzyme has less than 40% amino acid identity to FosE ([61] Figure 2). FruA of *S. mutans* plays a critical role in fructan degradation and the virulence of oral streptococci; BLAST analysis frequently identified homologues of FruA in other streptococci (Figure 2). Only five of the more than 1500 genome sequences assigned to the genus *Lactobacillus* harbors FruA homologues; this low frequency suggests that this β-fructanase is not necessary for the lifestyle of lactobacilli but only infrequently acquired by lateral gene transfer. Two of the species with FruA activity, *L. amylovorus* and *L. crispatus*, match species that are typically found in type II sourdoughs.

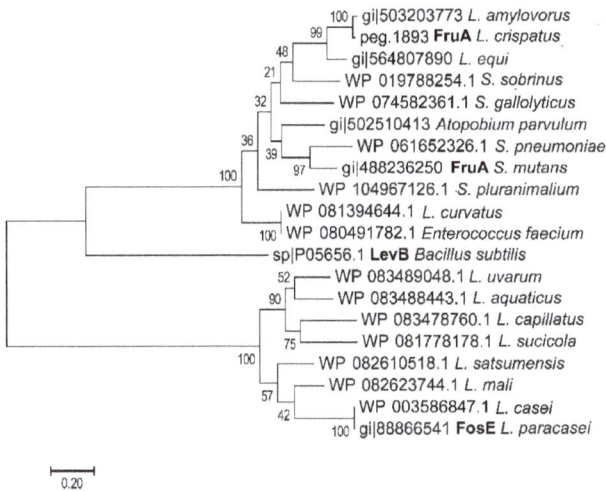

Figure 2. Molecular phylogenetic analysis of extracellular fructanases in lactic acid bacteria by the Maximum Likelihood method. The evolutionary history was inferred by using the Maximum Likelihood method; the tree is drawn to scale with branch lengths measured in the number of substitutions per site. Evolutionary analyses were conducted in MEGA7. Sequences were retrieved by NCBI Blast using the fructanase of *L. crispatus* [62] and the inulinase of *L. paracasei* [60] as query sequence. Sequences from lactic acid bacteria (*Lactobacillales*) with a more than 80% coverage and more than 50% amino acid identity were retrieved and aligned by ClustalW in MEGA 7.0. A levanase of *Bacillus subtilis* was included for comparison. Only one representative sequence for each bacterial species was chosen; sequences of 15 *Streptococcus* spp. which were all similar to sequences of other streptococci were omitted from the tree. The two *Lactobacillus* enzymes that were characterized biochemically are printed in bold.

Type I and type II sourdough microbiota generally include heterofermentative lactobacilli that convert fructose to mannitol. Degradation of mannitol in low FODMAP baking therefore requires mannitol-fermenting lactobacilli. Mannitol metabolism in lactobacilli is mediated by a mannitol-specific PTS system, followed by conversion by mannitol-1-phosphate-dehydrogenase to fructose-1-phospyate [63]. Enzymes for mannitol conversion are present in homofermentative lactobacilli of the *L. delbrueckii*, *L. casei*, *L. plantarum* and *L. salivarius* groups, likely representing trophic relationships with heterofermentative lactobacilli. In analogy to other PTS systems in lactobacilli, mannitol metabolism in homofermentative lactobacilli is repressed by glucose [39,41,58].

Glucose and maltose levels in wheat and rye sourdoughs and consequently carbon catabolite repression in homofermentative lactobacilli and yeasts [38,41] are determined by the level of damaged starch and the β-amylase and amyloglucosidase activity in flour (Figure 1; [18,64]). If enzyme activity and the level of damaged starch in flour are low, sucrose, raffinose, and fructans become the most readily available carbohydrates [64]. The composition of the microbiota in rye sourdoughs that are low in damaged starch match the composition in other type II sourdoughs with organisms of the *L. delbrueckii* group including *L. crispatus*, *L. amylovorus*, and *L. ultuensis*, and organisms of the *L. reuteri* group including *L. frumentii* and *L. pontis* as the dominant members [32,35,65]. The restricted availability of maltose and glucose, however, selects for strains expressing an exceptional fructanase (Figure 2, [62,65]). The prevailing enzyme activity is an extracellular exofructanase (Figure 2), which exhibits more than 80% of the maximum activity in the pH range of 4–6 and the temperature range of 30–60 °C [62]. Fructan hydrolysis in sourdough releases fructose that is partially converted to mannitol by *L. reuteri* group organisms (Figure 3). However, the restriction of carbohydrate sources also allows for mannitol conversion after fructans are completely consumed (Figure 3) and results in a virtually zero FODMAP sourdough. The use of this zero FODMAP sourdough in low FODMAP rye bread making involves the addition of unfermented rye flour, which is fermented for only a short time [65]. Nevertheless, the choice of appropriate raw materials and the use of FruA-positive and mannitol-fermenting lactobacilli allows fructan degradation in rye and rye sourdoughs for the production of bread with a low content of fructans and mannitol but a comparable fiber content when compared to regular bread [66–68].

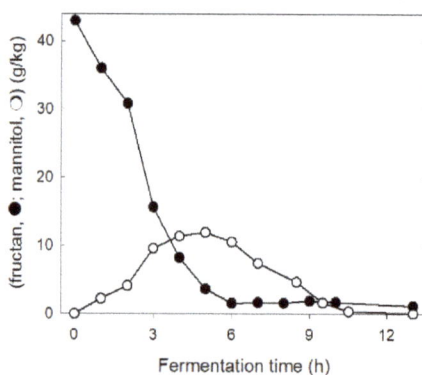

Figure 3. Degradation of fructans (black) and the formation and degradation of mannitol (white) in a type II rye sourdough. Sourdough microbiota consist of fructan-degrading strains of the *L. delbrueckii* group and heterofermentative strains of the *L. reuteri* group, which convert fructose to mannitol. Drawn with data from [65].

5. Proof of Concept from Clinical Trials with Low FODMAP Rye Bread

Two clinical trials done with IBS patients verified that low FODMAP rye bread made by using the above described zero FODMAP sourdough influences the gastrointestinal symptoms and the extent of gas production generated in intestinal fermentation. In the first study in a randomized double-blind controlled crossover study, it was shown that low FODMAP rye bread caused less flatulence, less abdominal pain, fewer cramps, and less stomach rumbling than regular rye bread [66]. Of note, the low FODMAP bread retained a high dietary fiber content (10 g/100 g) although the FODMAP levels were lowered to one third [66]. Including the low FODMAP rye bread thus also increased the dietary fiber intake to the recommended level in IBS patients, avoiding drawbacks of the other low FODMAP diets [15]. A second randomized double-blind controlled crossover study evaluated the amount of breath hydrogen levels after consuming low FODMAP rye bread or regular rye bread [68]. Low FODMAP rye bread reduced the generation of hydrogen by colonic fermentation [68]. This study showed that significant differences between bread types may occur in their postprandial effects.

6. Conclusions and Future Directions

Conventional sourdough baking reduces and converts FODMAPs in rye and wheat flour; however, the extent of FODMAP reduction is dependent on the fermentation organisms, the fermentation process, the grain raw material, and the sourdough dosage to the final bread dough. The production of low FODMAP bread requires extracellular fructanase activity; sourdough fermentation with lactobacilli expressing fructanases or the use of fructanase-positive yeasts provide wheat or rye breads with a low FODMAP content. Low FODMAP bread can help to restrict the intake of FODMAPs but at the same time increase the intake of slowly fermentable dietary fiber in IBS patients. High fiber/low FODMAP bread likely prevents the depletion of intestinal bifidobacteria that has been observed on other low FODMAP diets [14,15] and shows promise in reducing symptoms of IBS.

Anecdotal evidence links sourdough bread to improved tolerance of wheat in individuals with non-celiac wheat sensitivities [69]. In addition to the degradation of FODMAPs during sourdough fermentation, reduction and degradation of wheat amylase trypsin inhibitors may improve wheat tolerance in some individuals [67]. Amylase trypsin inhibitors are suggested to play a role in intestinal and extra-intestinal symptoms as they induce inflammatory reactions [70]. Amylase trypsin inhibitors are highly disulfide-bonded proteins; reduction of disulfide bonds reduces bioactivity and accelerates proteolytic digestion. Sourdough fermentation generates reducing conditions and supports reduction and hydrolysis of highly disulfide-bonded proteins that resist digestion in unfermented dough [71]. A pilot trial recruiting IBS patients with non-celiac wheat sensitivity, however, showed no improvement of intestinal symptoms after consuming sourdough wheat bread compared with industrial wheat bread [67]. Difficulties in identifying the protective effects of sourdough fermentation in non-celiac wheat intolerance relate to the poorly identified and likely multifactorial triggers of (self-diagnosed) non-celiac wheat sensitivity, and the inherent difficulties in blinding consumption of wheat or wheat sourdough products in clinical trials [67]. Despite the lack of support from clinical trials, sourdough-derived solutions likely play a significant role when developing healthier bakery products for people with non-gluten wheat sensitivities.

Author Contributions: Conceptualization, writing and editing: M.G.G. and J.L.

Funding: M.G.G. is supported by the Canada Research Chairs Program; J.L. is employed by a Fazer Group company Oy Karl Fazer Ab.

Conflicts of Interest: The authors declare no conflict of interest. The funding sponsors had no role in the writing of the manuscript, and in the decision to publish the review article.

References

1. Yan, Y.L.; Hu, Y.; Gänzle, M.G. Prebiotics, FODMAPs and dietary fibre–conflicting concepts in development of functional food products? *Curr. Opin. Food Sci.* **2018**, *20*, 30–37. [CrossRef]
2. Booijink, C.C.; El-Aidy, S.; Rajilić-Stojanović, M.; Heilig, H.G.; Troost, F.J.; Smidt, H.; Kleerebezem, M.; De Vos, W.M.; Zoetendal, E.G. High temporal and inter-individual variation detected in the human ileal microbiota. *Environ. Microbiol.* **2010**, *12*, 3213–3227. [CrossRef] [PubMed]
3. Zoetendal, E.G.; Raes, J.; van den Bogert, B.; Arumugam, M.; Booijink, C.C.; Troost, F.J.; Bork, P.; Wels, M.; de Vos, W.M.; Kleerebezem, M. The human small intestinal microbiota is driven by rapid uptake and conversion of simple carbohydrates. *ISME J.* **2012**, *6*, 1415–1426. [CrossRef] [PubMed]
4. Oku, T.; Nakamura, S. Digestion, absorption, fermentation, and metabolism of functional sugar substitutes and their available energy. *Pure Appl. Chem.* **2002**, *7*, 1253–1261. [CrossRef]
5. Oku, T.; Nakamura, S. Threshold for transitory diarrhea induced by ingestion of xylitol and lactitol in young male and female adults. *J. Nutr. Sci. Vitaminol. (Tokyo)* **2007**, *53*, 13–20. [CrossRef] [PubMed]

6. Murray, K.; Wilkinson-Smith, V.; Hoad, C.; Costigan, C.; Cox, E.; Lam, C.; Marciani, L.; Gowland, P.; Spiller, R.C. Differential effects of FODMAPs (fermentable oligo-, di-, mono-saccharides and polyols) on small and large intestinal contents in healthy subjects shown by MRI. *Am. J. Gastroenterol.* **2014**, *109*, 110–119. [CrossRef] [PubMed]

7. Hamaker, B.R.; Tuncil, Y.E. A perspective on the complexity of dietary fiber structures and their potential effect on the gut microbiota. *J. Mol. Biol.* **2014**, *426*, 3838–3850. [CrossRef] [PubMed]

8. Azcarate-Peril, M.A.; Ritter, A.J.; Savaiano, D.; Monteagudo-Mera, A.; Anderson, C.; Magness, S.T.; Klaenhammer, T.R. Impact of short-chain galactooligosaccharides on the gut microbiome of lactose-intolerant individuals. *Proc. Natl. Acad. Sci. USA* **2017**, *114*, E367–E375. [CrossRef] [PubMed]

9. Gerbault, P.; Liebert, A.; Itan, Y.; Powell, A.; Currat, M.; Burger, J.; Swallow, D.M.; Thomas, M.G. Evolution of lactase persistence: An example of human niche construction. *Philos. Trans. R. Soc. Lond. B Biol. Sci.* **2011**, *366*, 863–877. [CrossRef] [PubMed]

10. Latulippe, M.E.; Skoog, S.M. Fructose malabsorption and intolerance: Effects of fructose with and without simultaneous glucose ingestion. *Crit. Rev. Food Sci. Nutr.* **2011**, *51*, 583–592. [CrossRef] [PubMed]

11. Wilder-Smith, C.H.; Materna, A.; Wermelinger, C.; Schuler, J. Fructose and lactose intolerance and malabsorption testing: The relationship with symptoms in functional gastrointestinal disorders. *Aliment. Pharmacol. Ther.* **2013**, *37*, 1074–1183. [CrossRef] [PubMed]

12. Henström, M.; Diekmann, L.; Bonfiglio, F.; Hadizadeh, F.; Kuech, E.M.; von Köckritz-Blickwede, M.; Thingholm, L.B.; Zheng, T.; Assadi, G.; Dierks, C.; et al. Functional variants in the sucrase-isomaltase gene associate with increased risk of irritable bowel syndrome. *Gut* **2018**, *67*, 263–270. [CrossRef] [PubMed]

13. Bindels, L.B.; Delzenne, N.M.; Cani, P.D.; Walter, J. Towards a more comprehensive concept for prebiotics. *Nat. Rev. Gastroenterol. Hepatol.* **2015**, *12*, 303–310. [CrossRef] [PubMed]

14. Halmos, E.P.; Christophersen, C.T.; Bird, A.R.; Shepherd, S.J.; Gibson, P.R.; Muir, J.G. Diets that differ in their FODMAP content alter the colonic luminal microenvironment. *Gut* **2015**, *64*, 93–100. [CrossRef] [PubMed]

15. Staudacher, H.M.; Lomer, M.C.; Anderson, J.L.; Barrett, J.S.; Muir, J.G.; Irving, P.M.; Whelan, K. Fermentable carbohydrate restriction reduces luminal bifidobacteria and gastrointestinal symptoms in patients with irritable bowel syndrome. *J. Nutr.* **2012**, *142*, 1510–1518. [CrossRef] [PubMed]

16. Wilson, B.; Rossi, M.; Parkes, G.; Aziz, Q.; Anderson, W.; Irving, P.; Lomer, M.; Whelan, K. Prebiotic B-galacto-oligosaccharide supplementation of the low FODMAP diet improves symptoms of irritable bowel syndrome but does not prevent diet induced decline in bifidobacteria: A randomised controlled trial. *Proceed. Nutr. Soc.* **2017**, *76*. [CrossRef]

17. Campbell, J.M.; Bauer, L.L.; Fahey, G.C.; Hogarth, A.J.C.L.; Wolf, B.W.; Hunter, D.W. Selected fructooligosaccharide (1-kestose, nystose, and 1F-β-fructofuranosylnystose) composition of foods and feeds. *J. Agric. Food Chem.* **1997**, *45*, 3076–3082. [CrossRef]

18. Gänzle, M.G. Enzymatic and bacterial conversions during sourdough fermentation. *Food Microbiol.* **2014**, *37*, 2–10. [CrossRef] [PubMed]

19. De Giorgio, R.; Volta, U.; Gibson, P.R. Sensitivity to wheat, gluten and FODMAPs in IBS: Facts or fiction? *Gut* **2016**, *65*, 169–178. [CrossRef] [PubMed]

20. Schuppan, D.; Pickert, G.; Ashfaq-Khan, M.; Zevallos, V. Non-celiac wheat sensitivity: Differential diagnosis, triggers and implications. *Best Pract. Res. Clin. Gastroenterol.* **2015**, *29*, 469–476. [CrossRef] [PubMed]

21. Biesiekierski, J.R.; Peters, S.L.; Newnham, E.D.; Rosella, O.; Muir, J.G.; Gibson, P.R. No effects of gluten in patients with self-reported non-celiac gluten sensitivity after dietary reduction of fermentable, poorly absorbed, short-chain carbohydrates. *Gastroenterology* **2013**, *145*, 320–328. [CrossRef] [PubMed]

22. Verspreet, J.; Dorneza, E.; Van den Ende, W.; Delcour, C.; Courtin, C.M. Cereal grain fructans: Structure, variability and potential health effects. *Trends Food Sci. Technol.* **2015**, *43*, 32–42. [CrossRef]

23. Verspreet, J.; Pollet, A.; Cuyvers, S.; Vergauwen, R.; Van den Ende, W.; Delcour, J.A.; Courtin, C.M. A simple and accurate method for determining wheat grain fructan content and average degree of polymerization. *J. Agric. Food Chem.* **2012**, *60*, 2102–2107. [CrossRef] [PubMed]

24. Kuo, T.M.; Van Middlesworth, J.F.; Wolf, W.J. Content of raffinose oligosaccharides and sucrose in various plant seeds. *J. Agric. Food Chem.* **1988**, *36*, 32–36. [CrossRef]

25. Vinkx, C.J.A.; Delcour, J.A. Rye (*Secale cereale* L.) arabinoxylans: A critical review. *J. Cereal Sci.* **1996**, *24*, 1–14. [CrossRef]

26. Grausgruber, H.; Scheiblauer, J.; Schönlechner, R.; Ruckenbauer, P.; Berghofer, E. Variability in chemical composition and biologically active constituents of cereals. In *Genetic Variation for Plant Breeding*; Vollmann, J., Grausgruber, H., Ruckenbauer, P., Eds.; EUCARPIA & BOKU: Wien, Austria, 2004; pp. 23–26. ISBN 3-900962-56-1.

27. Brandt, M.J. Bedeutung von Rohwarenkomponenten. In *Handbuch Sauerteig*; Brandt, M.J., Gänzle, M.G., Eds.; Behr's Verlag: Hamburg, Germany, 2005; pp. 41–56. ISBN 3-89947-166-0.

28. Haskå, L.; Nymana, M.; Andersson, R. Distribution and characterisation of fructan in wheat milling fractions. *J. Cereal Sci.* **2008**, *48*, 768–774. [CrossRef]

29. Andersson, A.A.; Andersson, R.; Piironen, V.; Lampi, A.M.; Nyström, L.; Boros, D.; Fraś, A.; Gebruers, K.; Courtin, C.M.; Delcour, J.A.; et al. Contents of dietary fibre components and their relation to associated bioactive components in whole grain wheat samples from the HEALTHGRAIN diversity screen. *Food Chem.* **2013**, *136*, 1243–1248. [CrossRef] [PubMed]

30. Chateigner-Boutin, A.L.; Bouchet, B.; Alvarado, C.; Bakan, B.; Guillon, F. The wheat grain contains pectic domains exhibiting specific spatial and development-associated distribution. *PLoS ONE* **2014**, *9*, e89620. [CrossRef] [PubMed]

31. Saulnier, L.; Guillon, F.; Chateigner-Boutin, A.-L. Cell wall deposition and metabolism in wheat grain. *J. Cereal Sci.* **2012**, *56*, 91–108. [CrossRef]

32. Gänzle, M.; Ripari, V. Composition and function of sourdough microbiota: From ecological theory to bread quality. *Int. J. Food Microbiol.* **2016**, *239*, 19–25. [CrossRef] [PubMed]

33. Brandt, M.J.; Gänzle, M.G. *Handbuch Sauerteig*; Behr's Verlag: Hamburg, Germany, 2005; ISBN 3-89947-166-0.

34. Brandt, M.J. Sourdough products for convenient use in baking. *Food Microbiol.* **2007**, *24*, 161–164. [CrossRef] [PubMed]

35. De Vuyst, L.; Harth, H.; Van Kerrebroeck, S.; Leroy, F. Yeast diversity of sourdoughs and associated metabolic properties and functionalities. *Int. J. Food Microbiol.* **2016**, *239*, 26–34. [CrossRef] [PubMed]

36. Gobbetti, M.; Minervini, F.; Pontonio, E.; Di Cagno, R.; De Angelis, M. Drivers for the establishment and composition of the sourdough lactic acid bacteria biota. *Int. J. Food Microbiol.* **2016**, *239*, 3–18. [CrossRef] [PubMed]

37. Perlman, D.; Halvorson, H.O. Distinct repressible mRNAs for cytoplasmic and secreted yeast invertase are encoded by a single gene. *Cell* **1981**, *25*, 525–536. [CrossRef]

38. Gänzle, M.G.; Follador, R. Metabolism of oligosaccharides and starch in lactobacilli: A review. *Front. Microbiol.* **2012**, *3*, 340. [CrossRef] [PubMed]

39. Andersson, U.; Molenaar, D.; Radström, P.; de Vos, W.M. Unity in organization and regulation of catabolic operons in *Lactobacillus plantarum*, *Lactococcus lactis*, and *Listeria monocytogenes*. *Syst. Appl. Microbiol.* **2005**, *28*, 187–195. [CrossRef] [PubMed]

40. Teixeira, J.S.; Abdi, R.; Su, M.S.; Schwab, C.; Gänzle, M.G. Functional characterization of sucrose phosphorylase and scrR, a regulator of sucrose metabolism in *Lactobacillus reuteri*. *Food Microbiol.* **2013**, *36*, 432–439. [CrossRef] [PubMed]

41. Gänzle, M.G. Lactic metabolism revisited: Metabolism of lactic acid bacteria in food fermentations and food biotechnology. *Curr. Opin. Food Sci.* **2015**, *2*, 106–117. [CrossRef]

42. Gänzle, M.G.; Vermeulen, N.; Vogel, R.F. Carbohydrate, peptide and lipid metabolism of lactic acid bacteria in sourdough. *Food Microbiol.* **2007**, *24*, 128–138. [CrossRef] [PubMed]

43. Zheng, J.; Ruan, L.; Sun, M.; Gänzle, M. A genomic view of lactobacilli and pediococci demonstrates that phylogeny matches ecology and physiology. *Appl. Environ. Microbiol.* **2015**, *81*, 7233–7243. [CrossRef] [PubMed]

44. Galle, S.; Arendt, E.K. Exopolysaccharides from sourdough lactic acid bacteria. *Crit. Rev. Food Sci. Nutr.* **2014**, *54*, 891–901. [CrossRef] [PubMed]

45. Galle, S.; Schwab, C.; Arendt, E.; Gänzle, M. Exopolysaccharide-forming *Weissella* strains as starter cultures for sorghum and wheat sourdoughs. *J. Agric. Food Chem.* **2010**, *58*, 5834–5841. [CrossRef] [PubMed]

46. Schwab, C.; Mastrangelo, M.; Corsetti, A.; Gänzle, M.G. Formation of oligosaccharides and polysaccharides by *Lactobacillus reuteri* LTH5448 and *Weissella cibaria* 10M in sorghum sourdoughs. *Cereal Chem.* **2008**, *85*, 679–684. [CrossRef]

47. Korakli, M.; Rossmann, A.; Gänzle, M.G.; Vogel, R.F. Sucrose metabolism and exopolysaccharide production in wheat and rye sourdoughs by *Lactobacillus sanfranciscensis*. *J. Agric. Food Chem.* **2001**, *49*, 5194–5200. [CrossRef] [PubMed]

48. Teixeira, J.S.; McNeill, V.; Gänzle, M.G. Levansucrase and sucrose phoshorylase contribute to raffinose, stachyose, and verbascose metabolism by lactobacilli. *Food Microbiol.* **2012**, *31*, 278–284. [CrossRef] [PubMed]

49. Zhao, X.; Gänzle, M.G. Genetic and phenotypic analysis of carbohydrate metabolism and transport in *Lactobacillus reuteri*. *Int. J. Food Microbiol.* **2018**, *272*, 12–21. [CrossRef] [PubMed]

50. Ostergaard, S.; Olsson, L.; Nielsen, J. Metabolic Engineering of *Saccharomyces cerevisiae*. *Microbiol. Mol. Biol. Rev.* **2000**, *64*, 34–50. [CrossRef] [PubMed]

51. Whelan, K.; Abrahmsohn, O.; David, G.J.; Staudacher, H.; Irving, P.; Lomer, M.C.; Ellis, P.R. Fructan content of commonly consumed wheat, rye and gluten-free breads. *Int. J. Food Sci. Nutr.* **2011**, *62*, 498–503. [CrossRef] [PubMed]

52. Brandt, J.J.; Hammes, W.P. Einfluss von Fructosanen auf die Sauerteigfermentation. *Getreide Mehl Brot* **2001**, *55*, 341–345.

53. Struyf, N.; Laurent, J.; Verspreet, J.; Verstrepen, K.J.; Courtin, C.M. *Saccharomyces cerevisiae* and *Kluyveromyces marxianus* co-cultures allow reduction of fermentable oligo-, di-, and monosaccharides and polyols levels in whole wheat bread. *J. Agric. Food Chem.* **2017**, *65*, 8704–8713. [CrossRef] [PubMed]

54. Nilsson, U.; Öste, R.; Jägerstad, M. Cereal fructans: Hydrolysis by yeast invertase, in vitro and during fermentation. *J. Cereal Sci.* **1987**, *6*, 53–60. [CrossRef]

55. Sainz-Polo, M.A.; Ramírez-Escudero, M.; Lafraya, A.; González, B.; Marín-Navarro, J.; Polaina, J.; Sanz-Aparicio, J. Three-dimensional structure of *Saccharomyces* invertase: Role of a non-catalytic domain in oligomerization and substrate specificity. *J. Biol. Chem.* **2013**, *288*, 9755–9766. [CrossRef] [PubMed]

56. Kaplan, H.; Hutkins, R.W. Metabolism of fructooligosaccharides by *Lactobacillus paracasei* 1195. *Appl. Environ. Microbiol.* **2003**, *69*, 2217–2222. [CrossRef] [PubMed]

57. Saulnier, D.M.; Molenaar, D.; de Vos, W.M.; Gibson, G.R.; Kolida, S. Identification of prebiotic fructooligosaccharide metabolism in *Lactobacillus plantarum* WCFS1 through microarrays. *Appl. Environ. Microbiol.* **2007**, *73*, 1753–1765. [CrossRef] [PubMed]

58. Barrangou, R.; Azcarate-Peril, M.A.; Duong, T.; Conners, S.B.; Kelly, R.M.; Klaenhammer, T.R. Global analysis of carbohydrate utilization by *Lactobacillus acidophilus* using cDNA microarrays. *Proc. Natl. Acad. Sci. USA* **2006**, *103*, 3816–3821. [CrossRef] [PubMed]

59. Stuyf, N.; Vancdewiele, H.; Herrera-Malaver, B.; Verspreet, J.; Verstrepen, K.J.; Courtin, C.M. *Kluyveromyces marxianus* yeast enables the production of low FODMAP whole wheat breads. *Food Microbiol.* **2018**, *76*, 135–145. [CrossRef]

60. Goh, Y.J.; Lee, J.H.; Hutkins, R.W. Functional analysis of the fructooligosaccharide utilization operon in *Lactobacillus paracasei* 1195. *Appl. Environ. Microbiol.* **2007**, *73*, 5716–5724. [CrossRef] [PubMed]

61. Burne, R.A.; Penders, J.E. Differential localization of the *Streptococcus mutans* GS-5 fructan hydrolase enzyme, FruA. *FEMS Microbiol. Lett.* **1994**, *121*, 243–249. [CrossRef] [PubMed]

62. Loponen, J.; Mikola, M.; Sibakov, J. An Enzyme Exhibiting Fructan Hydrolase Activity. Patent No. WO2017220864A1, 28 December 2017.

63. Wisselink, H.W.; Moers, A.P.; Mars, A.E.; Hoefnagel, M.H.; de Vos, W.M.; Hugenholtz, J. Overproduction of heterologous mannitol 1-phosphatase: A key factor for engineering mannitol production by *Lactococcus lactis*. *Appl. Environ. Microbiol.* **2005**, *71*, 1507–1514. [CrossRef] [PubMed]

64. Struyf, N.; Laurent, J.; Lefevere, B.; Verspreet, J.; Verstrepen, K.J.; Courtin, C.M. Establishing the relative importance of damaged starch and fructan as sources of fermentable sugars in wheat flour and whole meal bread dough fermentations. *Food Chem.* **2017**, *218*, 89–98. [CrossRef] [PubMed]

65. Loponen, J. Low-Fructan Grain Material and a Method for Producing the Same. Patent No. WO2016113465A1, 21 July 2016.

66. Laatikainen, R.; Koskenpato, J.; Hongisto, S.M.; Loponen, J.; Poussa, T.; Hillilä, M.; Korpela, R. Randomised clinical trial: Low-FODMAP rye bread vs. regular rye bread to relieve the symptoms of irritable bowel syndrome. *Aliment. Pharmacol. Ther.* **2016**, *44*, 460–470. [CrossRef] [PubMed]

67. Laatikainen, R.; Koskenpato, J.; Hongisto, S.M.; Loponen, J.; Poussa, T.; Huang, X.; Sontag-Strohm, T.; Salmenkari, H.; Korpela, R. Pilot study: Comparison of sourdough wheat bread and yeast-fermented wheat bread in individuals with wheat sensitivity and irritable bowel syndrome. *Nutrients* **2017**, *9*, E1215. [CrossRef] [PubMed]

68. Pirkola, L.; Laatikainen, R.; Loponen, J.; Hongisto, S.M.; Hillilä, M.; Nuora, A.; Yang, B.; Linderborg, K.M.; Freese, R. Low-FODMAP vs regular rye bread in irritable bowel syndrome: Randomized SmartPill® study. *World J. Gastroenterol.* **2018**, *24*, 1259–1268. [CrossRef] [PubMed]

69. CBC. 2013. Available online: http://www.cbc.ca/news/health/sourdough-breadmaking-cuts-gluten-content-in-baked-goods-1.2420209 (accessed on 26 May 2018).

70. Junker, Y.; Zeissig, S.; Kim, S.J.; Barisani, D.; Wieser, H.; Leffler, D.A.; Zevallos, V.; Libermann, T.A.; Dillon, S.; Freitag, T.L.; et al. Wheat amylase trypsin inhibitors drive intestinal inflammation via activation of toll-like receptor 4. *J. Exp. Med.* **2012**, *209*, 2395–2408. [CrossRef] [PubMed]

71. Loponen, J.; König, K.; Wu, J.; Gänzle, M.G. Influence of thiol metabolism of lactobacilli on egg white proteins in wheat sourdoughs. *J. Agric. Food Chem.* **2008**, *56*, 3357–3362. [CrossRef] [PubMed]

MDPI

St. Alban-Anlage 66

4052 Basel

Switzerland

Tel. +41 61 683 77 34

Fax +41 61 302 89 18

www.mdpi.com

Foods Editorial Office

E-mail: foods@mdpi.com

www.mdpi.com/journal/foods